化学工业出版社"十四五"普通高等教育本科规划教材

Organic Chemistry Experiments

有机化学实验

（双语版）

王 红　梁仁校　强根荣　主编

内容简介

《有机化学实验（双语版）》全书共 6 章，分别为：有机化学实验的一般知识、有机化学实验的基本操作、有机化学基础性实验、有机化学综合性实验、研究性文献实验和典型实验的学习指导。本书在立足基本能力培养的同时，还补充完善了新型的有机合成技术和综合性实验内容，以实现知识、能力的传承与创新协调发展。结合双语实验教学的特点和存在的问题，本书以英文为主，中文为辅，以期给读者提供一个相对完整的专业英语的学习环境，提升其专业英语的阅读、理解和书写能力。

本书可作为化学、化学工程与工艺、应用化学、能源化学工程、安全工程、材料科学与工程、生物制药、环境科学等专业的双语教学配套实验教材，也可供留学生和相关专业人员参考。

图书在版编目（CIP）数据

有机化学实验：双语版：汉、英/王红，梁仁校，强根荣主编 .—北京：化学工业出版社，2024.3
ISBN 978-7-122-44551-3

Ⅰ.①有… Ⅱ.①王…②梁…③强… Ⅲ.①有机化学-化学实验-高等学校-教材-汉、英 Ⅳ.①O62-33

中国国家版本馆 CIP 数据核字（2023）第 232763 号

责任编辑：汪 靓 宋林青　　装帧设计：史利平
责任校对：王 静

出版发行：化学工业出版社
　　　　　（北京市东城区青年湖南街 13 号　邮政编码 100011）
印　　装：三河市延风印装有限公司
787mm×1092mm　1/16　印张 15½　字数 376 千字
2024 年 3 月北京第 1 版第 1 次印刷

购书咨询：010-64518888　　　售后服务：010-64518899
网　　址：http://www.cip.com.cn
凡购买本书，如有缺损质量问题，本社销售中心负责调换。

定　价：39.80 元　　　　　　　　版权所有　违者必究

Foreword

Organic chemistry experiment is an important basic course for chemistry, chemical engineering, pharmacy and other related majors. Bilingual or full English teaching is widely used in the teaching of organic chemistry experiments today, which can help students acquire chemistry knowledge and improve experimental skills, meanwhile, expand the space for creating and using the second language in chemistry learning, and establish cross-cultural awareness and develop the ability of cross-culture communication.

Zhejiang University of Technology has carried out bilingual teaching of organic chemistry experiments for more than ten years, and has been using self-compiled full-English teaching materials. With the continuous development of organic chemistry theory and experimental technology, the bilingual course of organic chemistry experiment has undergone great changes in its concept, content and methods. Previous teaching materials for organic chemistry experiments no longer suits the need for today's bilingual course. Therefore, on the basis of previous teaching practices and referring to the newly published organic chemistry experiment textbooks worldwide, we modified the content of the original teaching materials, added new technologies for organic chemistry experiments, and completed *Organic Chemistry Experiment* (*bilingual edition*).

This book consists of 6 chapters, including 15 important organic experiment units, 32 organic chemistry experiments ranging from the fundamental to rather comprehensive ones, 10 research literature experiments, and study guidance for 9 typical experiments. Based on the reference to traditional domestic bilingual organic chemistry experiment textbooks, this book has innovated the design of organic chemistry experiment materials in following aspects:

1. The arrangement of the content is more challenging and systematic

This book no longer adopts the traditional half English-half Chinese translation format of bilingual textbooks. Instead, it is predominantly in English. Only some key concepts are supplemented with Chinese translation and hints in the notes of each page. This arrangement provides students with a relatively consistent professional English learning environment to reduce students' dependence on Chinese learning and switch them to actively learn professional English.

In order to effectively connect organic chemistry theory and experiment, facilitate understanding of the reaction mechanism, and clarify the purpose of each experiment step. Chapter 3 of "Fundamental Experiments" was written in the logic of the types of functional groups contained in the target compounds. We put the synthesis of compounds containing the same functional group in the same section. Such arrangement makes the book clearer and more systematic, which is basically consistent with the theoretical study of organic chemistry.

2. The content of teaching materials is more comprehensive and novel

(1) Emphasize laboratory safety and enhance students' awareness of environmental protection.

In Chapter 1 of "Introduction of Experimental Organic Chemistry", the green chemistry and laboratory waste treatment methods are added to reflect the concept of green and environmental protection. In addition, by referring to the MSDS (Material Safety Data Sheet), students are provided with "help hints" for the chemical reagents used in the synthesis experiment in terms of the danger of the reagents, personal protection, first-aid measures, and handling.

(2) Focus on the frontier and reflect the new technology of organic chemistry experiments.

In Chapter 4 of "Comprehensive Experiments", some experiments that can be safely operated in undergraduate laboratory and reflect new synthetic techniques are added, such as microwave reaction, solvent-free reaction, combined reaction, etc. Most of the comprehensive experiments mentioned in this book are multi-step reactions. Some experiments also connect synthesis, separation, purification and characterization of target products to enhance the separation and purification operations of products, which effectively strengthen the whole process training for students in organic synthesis.

Chapter 5 of "Experiments from Research and Literature" introduces some experiments that are not yet available for undergraduate laboratory due to experimental environment restrictions, but can well reflect the latest achievements of modern organic synthesis, such as water-free and air-free reactions, solvent-free grinding reactions, transition metal catalyzed reactions, electronic reactions, ultrasonic assisted reactions, green synthetic reactions, etc. These experiments are aimed to broaden students' horizons and cultivate students' innovative consciousness and ability. Students' professional English reading and writing skills will also be improved in exploring and reading those literature.

(3) Enrich the learning resources of students

The Chapter 6 of "Study Guide of Some Typical Experiments" are extracurricular learning materials for some of the experiments taught in Zhejiang University of Technology. This chapter includes introduction, techniques involved, safety alerts, pre-lab assignment and thinking questions, etc. for 9 typical experiments. These contents can be used to examine the preview and learning outcomes, and help students better summarize and command knowledge covered by experiments.

This book was written by Hong Wang, Renxiao Liang, and Genrong Qiang. The whole book was consolidated by Wang Hong. Ling Liu, Hongwei Jin, Xiaoliang Xu and other teachers of the organic chemistry experiment teaching team also contribute many useful suggestions. This textbook was funded by the key textbook construction project of Zhejiang University of Technology. We have also referred to many domestic and foreign organic chemistry experiment textbooks and related documents during the writing process. Here we would like to express our heartfelt thanks as well.

Last but not least, we well recognize that imperfectness and errors cannot fully avoided. Comments, thoughts, and criticism that help polish the book will be much welcomed and valued.

Editor
May, 2023 in Hangzhou

前言

有机化学实验是化学、化工、制药及其他近化学类专业一门重要的基础课程。在有机化学实验教学中使用双语或全英文授课，可以帮助学生在获得学科知识、提高实验能力的同时，拓展学习和使用第二语言的空间，培养跨文化交流的能力。

浙江工业大学开展有机化学实验双语教学已有十多年，教学中一直使用自编的全英文实验讲义。随着有机化学理论与实验技术的不断发展，有机化学实验双语课程在教学理念、教学内容和教学方式等方面都有了很大的变化，原有的实验讲义已不能满足现代有机化学实验双语教学的要求。为此，我们在总结有机化学实验双语教学经验的基础上，参考了近年来出版的国内外有机化学实验教材，对原有讲义的内容进行了修订，增补了有机化学实验的新理念、新技术，编写完成了《有机化学实验》（双语版）教材。

全书共6章，包括有机化学的一般知识、15个重要的有机实验基本操作、32个从基础性到综合性的有机化学实验、10个研究性文献实验以及9个典型实验的学习指导。本书在参考国内传统有机化学实验双语教材的基础上，在编排方式和内容设置等方面进行了以下几个方面新的尝试。

1. 编排方式更具挑战性、系统性

本书内容的编写不再采用双语教材传统的一半英文、一半中文翻译的形式，而是以全英文为主，只对一些重点、难点内容以注释的形式辅以中文翻译和提示。这样的编排给学生提供了一个相对完整的专业英语的学习环境，以减少学生中文学习的依赖性，引导学生主动进行专业英语的学习。

为了加强有机化学理论课程与实验课程的联系，便于学生理解实验所涉及的反应机理，明确每一步实验操作的目的，本书第3章"有机化学基础实验"以目标化合物所含官能团种类为主线进行编写，把含有相同官能团的化合物的合成放在同一节中。这样的编排使内容更为清晰和系统，并与有机化学理论学习基本保持了一致。

2. 教材内容更全面、更具新颖性

（1）充分强调实验室安全，提升学生的环保意识

在第1章"有机化学实验的一般知识"中，增加了绿色化学和实验室废物处理方法等体现绿色、环保理念的内容。另外，还通过查阅MSDS（Material Safety Data Sheet），对合成实验中用到的化学试剂，从试剂危险性、个人防护、急救措施和操作处置等方面为学生提供"help hints"。

（2）实验内容关注前沿，体现有机化学实验新技术

在第4章"综合性实验"中，增加了一些在本科实验教学中能安全运行，并且能体现有机化学新型合成技术的实验内容，如：微波反应、无溶剂反应、组合式反应等。综合实验中的大部分实验为多步反应，部分实验还将合成、分离、提纯以及目标产物的表征串联在一

起，强化产物的分离和提纯操作，加强对学生有机合成全过程的训练。

第 5 章"研究性文献实验"，向学生介绍一些在本科实验教学中尚不具备条件开设，但能体现现代有机合成部分发展成果的内容，如无水无氧反应、无溶剂研磨反应、过渡金属催化反应、有机光反应、超声波辐射反应、绿色合成反应等。这些文献实验旨在拓展学生的视野，培养学生的创新意识和能力。同时，学生通过相关英文文献的查阅学习，其专业英语的阅读、理解和书写能力也会得到一定提升。

（3）丰富学生的学习资源

本书的第 6 章"典型实验的学习指导"是针对我校开设的部分实验内容，给学生提供的课外学习资料，内容包括 9 个典型实验的实验原理、实验技术、安全提示、课前预习和课后拓展思考题目等。这些内容可以用于学生对预习情况和学习效果的自查，帮助学生更好地总结和消化相关的实验知识。

本书由王红、梁仁校、强根荣分工编写完成，全书由王红统稿。刘玲、金红卫、许孝良老师和有机化学实验教学团队的其他老师给本书的编写提出了很多有益的建议。本教材得到了浙江工业大学重点教材建设项目的资助。在编写过程中，参考了许多国内外有机化学实验教材和相关的文献资料，在此一并表示衷心的感谢。

限于作者水平，书中不当之处在所难免，敬请使用者批评指正。

编者

2023 年 5 月于杭州

Chapter 1 — 1
Introduction of Experimental Organic Chemistry

- 1.1 The Role of Experimental Organic Chemistry 1
- 1.2 General Rules for the Organic Chemistry Experiments 2
- 1.3 Safety in the Organic Chemistry Laboratory 3
- 1.4 Pre-lab Reports, Notebooks and Lab Reports in the Laboratory 8
- 1.5 Common Laboratory Glassware and Equipment 15
- 1.6 The Literature of Organic Chemistry 18
- 1.7 Green Chemistry in the Organic Laboratory 21

Chapter 2 — 25
Basic Operations of Organic Chemistry Experiments

- 2.1 Heating and Cooling 25
- 2.2 Drying Organic Solutions and Solids 28
- 2.3 Filtration 31
- 2.4 Simple Distillation and Boiling Point 33
- 2.5 Fractional Distillation 36
- 2.6 Steam Distillation 38
- 2.7 Vacuum Distillation 41
- 2.8 Extraction and Washing 44
- 2.9 Recrystallization 48
- 2.10 Sublimation 53
- 2.11 Measurement of the Melting Point 55
- 2.12 Refluxing 58
- 2.13 Stirring Methods 61
- 2.14 Water-free and Air-free Operation 62
- 2.15 Chromatography Techniques 64

Chapter 3
Fundamental Experiments — 80

- 3.1 Unsaturated Hydrocarbons ······ 80
  - Exp. 1 Preparation of Cyclohexene ······ 80
  - Exp. 2 Preparation of Phenylacetylene ······ 82
- 3.2 Alkyl Halides ······ 85
  - Exp. 3 Preparation of 1-Bromobutane ······ 85
  - Exp. 4 Preparation of 2-Chloro-2-methylbutane ······ 89
- 3.3 Alcohols, Phenols and Ethers ······ 92
  - Exp. 5 Preparation of 2-Methyl-2-butanol ······ 92
  - Exp. 6 Preparation of Triphenylmethanol ······ 96
  - Exp. 7 Preparation of Di-n-butyl Ether ······ 100
  - Exp. 8 Preparation of Ethyl Phenyl Ether ······ 103
  - Exp. 9 Preparation of o-tert-Butylhydroquinone ······ 106
- 3.4 Aldehydes and Ketones ······ 109
  - Exp. 10 Preparation of Cyclohexanone ······ 109
  - Exp. 11 Preparation of Acetophenone ······ 112
  - Exp. 12 Preparation of Benzaldehyde ······ 115
- 3.5 Carboxylic Acids ······ 118
  - Exp. 13 Preparation of Benzoic Acid ······ 118
  - Exp. 14 Preparation of Cinnamic Acid ······ 121
  - Exp. 15 Preparation of Hexane-1,6-dionic Acid ······ 124
- 3.6 Carboxylic Acid Derivatives ······ 127
  - Exp. 16 Preparation of Ethyl Acetate ······ 127
  - Exp. 17 Preparation of n-Dibutyl Phthalate ······ 130
- 3.7 Nitrogen-containing Compounds ······ 133
  - Exp. 18 Preparation of Acetanilide ······ 133
  - Exp. 19 Preparation of Methyl Orange ······ 135
- 3.8 Heterocyclic Compounds ······ 138
  - Exp. 20 Preparation of 8-Hydroxyquinoline ······ 138
  - Exp. 21 Extraction and Separation of Caffeine from Tea Leaves ······ 141

Chapter 4
Comprehensive Experiments — 144

- Exp. 1 Preparation of Benzyl Alcohol and Benzoic Acid ······ 144
- Exp. 2 Preparation of Benzil and Thin-layer Chromatography ······ 147
- Exp. 3 Preparation of Acetylferrocene and Column Chromatography ······ 150
- Exp. 4 Preparation of Ethyl Acetoacetate by the Claisen Condensation

		Reaction	154
	Exp. 5	Synthesis and Characterization of Aspirin	157
	Exp. 6	Multi-step Synthesis of Sulfanilamide	161
	Exp. 7	A Solvent Free Cannizzaro Reaction and Thin-layer Chromatography	166
	Exp. 8	Preparation of 4-Vinylbenzoic Acid by a Wittig Reaction	169
	Exp. 9	Preparation of Dimedone via the Robinson Annulation	172
	Exp. 10	Preparation of Cinnamic Acid by the Knoevenagel Condensation	175
	Exp. 11	Microwave-assisted Preparation of Benzoic Acid and Ethyl Benzoate	178

Chapter 5 183
Experiments from Research and Literature

	Exp. 1	Synthesis of 3-Substituted Isoindolin-1-ones by Solvent-free Grinding Reaction	183
	Exp. 2	Electroorganic Synthesis of Nitriles	185
	Exp. 3	Ultrasonic Assisted Green Protocol for the Synthesis of N,N′-bis Phenylsulfamide	187
	Exp. 4	Clean Synthesis in Water: Uncatalysed Preparation of Ylidenemalononitriles	188
	Exp. 5	Transition-metal-catalyzed Convenient Synthesis of Internal Alkyne	190
	Exp. 6	N-Heterocyclic Carbene Catalyzed Intramolecular Benzoin Condensation to Access 9,10-Phenanthraquinone	192
	Exp. 7	Optical Resolution of (±)-(S)-Methyl-(S)-phenylsulfoximine with (+)-L-Camphorsulfonic Acid	194
	Exp. 8	Synthesis of Benzyl Acetate in Room Temperature Ionic Liquids	197
	Exp. 9	Green Synthesis of 4-Bromoaniline	200
	Exp. 10	Pinacol Rearrangement and Photochemical Synthesis of Benzopinacol	204

Chapter 6 207
Study Guide of Some Typical Experiments

	Exp. 1	Preparation of Cyclohexene (Miniscale Reaction)	207
	Exp. 2	Preparation of 1-Bromobutane	209
	Exp. 3	Preparation of Acetyl Ferrocene and Column Chromatography	212
	Exp. 4	Preparation of Acetylaniline	215
	Exp. 5	Preparation of 2-Methyl-2-butanol	217
	Exp. 6	Preparation of Benzil and Thin-layer Chromatography	220

Exp. 7　Preparation of Cinnamic Acid ·· 223
Exp. 8　Preparation of Benzoic Acid and Ethyl Benzoate ······················ 225
Exp. 9　Preparation of Benzoic Acid and Benzyl Alcohol ······················ 229

Appendix —————————————————————————— 232

References ————————————————————————— 235

Chapter 1
Introduction of Experimental Organic Chemistry

1.1 The Role of Experimental Organic Chemistry

Experimental organic chemistry is an integral part of organic chemistry course and plays an important role in developing and reinforcing your understanding what your organic chemistry textbook presents. It also provides you an opportunity to test and verify the principles that you have learned from organic chemistry. The abstract theoretical concepts, the confusing phenomena and the mechanisms of the reactions mentioned in class will be more understandable through experiments in the laboratory.

The basic goal of experimental organic chemistry is to introduce the techniques and procedures for carrying out organic chemical reactions, separating and purifying the desired products from the reaction mixtures. In order to accomplish the above tasks better, you have to learn how to handle chemicals safely, how to assembe apparatus properly, how to observe and record experimental phenomena, and how to obtain accurate experimental results. All of these are the bases for you to have the successful practice in organic chemistry laboratory and future research. Meanwhile, you should also develop the abilities to design experiments, to interpret observed phenomena and experimental results, and to draw reasonable conclusions from experiments, which are the heart of doing research. In organic chemistry laboratory, there will also be some risks and hazardous chemicals in some experiments. Green chemistry consciousness and environmental protection idea are required to minimize the risks and to protect the environment.

The work in the undergraduate organic chemistry laboratory is just the beginning of your scientific journey. In addition to the basic experimental techniques, the cultivation of the qualities such as serious attitude, active participation, creativity and critical thinking are also of great benefit for your future work and research. The organic chemistry laboratory is where you learn about "how we know what we know about it". You should develop good habits and follow the process of science—to observe, to think, and to act.

1.2 General Rules for the Organic Chemistry Experiments

In order to ensure that the experiments go smoothly and effectively, and to protect the safety of individual students and laboratories, all students must abide the following general rules during any organic experimental procedure:

① You must know what and where the safety features are and know how to operate them. The safety features in organic laboratories include fire extinguishers and sand bucket, safety showers, eye wash stations, and first aid kits, etc❶. It is of vital importance to be crystal clear about what to do in danger in the process of experiments.

② Become knowledgeable about basic first-aid procedures. The damage from accidents will be minimized if first aid is applied promptly.

③ Check all experimental instruments and equipment carefully, then inspect whether water, power and gas supply are functioning well. If there is abnormal condition, report to your instructor immediately.

④ You're not allowed to enter into the lab without wearing a lab coat. Slippers and skirts are prohibited in the lab to prevent physical harm. Eating is also forbidden in organic chemistry lab. Drinking is allowed in some designated place only.

⑤ Good planning and previewing are required before starting an experiment. You should plan your work ahead of time and be clear about the objectives and procedures of the experiment. You should accomplish the pre-lab report❷ in advance.

⑥ Be familiar with the property and safe usage of the chemicals to be used in the experiment. You should wear appropriate gloves and safety goggles when handling with the chemical reagents. Transfer only the needed amount of reagents out and never return unused reagent back to its container❸. Please put the commonly used reagents in the designated place and don't change the fixed position randomly.

⑦ Wipe off the reagent bottles and replace the caps as soon as possible after measuring the desired chemical reagents. This can not only prevent the volatilization of the reagents, but also avoid the contamination of the reagents caused by using the wrong caps. Clean the spills on the floor or bench immediately and never leave a mess for others.

⑧ Carry out the reaction following the experimental produces. As you perform each step, think about why you are doing this particular step. Think carefully before moving to the next step. If you want to make any changes, please communicate with your instructor and get authorization from him or her first.

⑨ You should be concentrated on what you are doing, operate with care, keep close observation, and record the experimental phenomenon precisely. Horseplay and running are

---

❶ 有机化学实验室常用的安全设施包括：灭火器、沙桶、安全淋浴、洗眼池和急救箱等。
❷ 实验预习报告。
❸ 按所需量量取药品，不可将未用完的药品重新倒回试剂瓶。

dangerous and should be strictly forbidden in the laboratory.

⑩ Keep your laboratory space clean and neat. Put the rubbish in the designated place, and place the liquids and solids waste in various specified containers.

⑪ Clean all the used glassware and your bench area by the end. Make sure that all the gas, water and power valves are closed.

⑫ Wash your hands with soap and water and ask the instructor for permission before leaving the lab.

Overall, in order to achieve a successful laboratory experience, you should always follow safety rules, plan your laboratory work in advance, operate with care and patience, and keep a detailed record of the experiment.

1.3 Safety in the Organic Chemistry Laboratory

1. General Safety Rule for an Undergraduate Laboratory

Organic chemistry experiments commonly involve the use of flammable and toxic chemicals, fragile glassware, electronic equipment and gas cylinders, etc. All of them entail potential risks. Thus, it is crucial to prevent accidents. When you start your work in organic laboratory, you should not only have a comprehensive understanding of safety principles but also be able to put those principles into practice.

The accidents occurred in the organic laboratory are generally of three types: fires or explosions, cuts or burns, and accidents occurring from inhalation, absorption or ingestion of toxic materials. The information involved in this section is the instruction to prevent those accidents.

(1) Fire-proof[4]

Fire is the chemical union of flammable materials with oxygen in the air. Fires can be prevented in the laboratory by keeping flammable chemicals away from flame sources. There are three kinds of flame sources in the organic laboratory: open flames[5], hot surfaces such as hot plates or heating mantles[6], and electrical equipment.

① Open flames

Most organic substances are combustible, and some are highly volatile as well, therefore,

- Never use open flames to heat organic compounds.

- Do not transfer flammable liquids from one container to another when there are open flames nearby.

- Check the glassware carefully before the experiment begins. Heating a flask having "star" cracks may lead to the fracture of flask, which will result in a serious fire caused by the combustion of leaked chemicals. In fact, the use of an open flame in the organic laboratory

---

[4] 防火。

[5] 明火。

[6] 电热套。

should be forbidden and done only with the permission of your instructor.

② Hot surfaces

A flammable liquid spilled or heated on a hot surface, such as a hot plate or a heating mantle may burst into flame. This is because the vapors of the flammable liquid can be ignited by the hot surfaces. Therefore,

- Do not heat any volatile liquid in an open container[7] on hot surfaces.
- Equip the liquid container with a reflux condenser[8] if a volatile liquid must be heated on hot surfaces.
- Remove the hot heating mantle or hot plate away before transferring the volatile organic liquid.

③ Electrical equipment

The vapors of the flammable liquid can be ignited by a spark from the damaged electrical equipment. Thus, you should avoid using any instrument with frayed or faulty electrical equipment to prevent electrical fire. Make sure that there are no frayed electrical cords before using electrical equipment.

In case of fire, you should be calm and try to take proper measures to control the expansion of the accident. If the situation is manageable, first of all, cut off the power and move the combustibles and explosive substances in the vicinity to a safe place. Then, choose the appropriate measures based on the nature of the combustible substance and the size of the flame to extinguish the fire. In general, the fire caused by the combustion of organic substances can not be put out with water. A small fire can be extinguished with asbestos cloth. When the fire is bigger, use a fire extinguisher to put it out. When the fire spreads on the surface of your bench or on the ground, cover the surface with sand to put the fire out. If the fire is out of control, call 119, inform your instructor immediately, and leave the laboratory as soon as possible.

(2) **Explosion-proof**[9]

In the organic chemistry laboratory, there are two common kinds of explosion according to the nature of the explosions:

① Chemicals explosion

Some organic substances are explosive, such as peroxides[10] and trinitrotoluene[11], etc. When these substances are heated or hit, they might explode easily.

② Instruments explosion

Incorrect installation or improper operation of experimental instruments can also cause explosions. For example, the whole set of experimental apparatus is airtight in the process of

---

[7] 敞口容器。

[8] 回流冷凝管。

[9] 防爆。

[10] 过氧化物。

[11] 三硝基甲苯。

simple distillation, or an Erlenmeyer flask is used as the receiving container during vacuum distillation, etc.

In order to prevent the explosion accident, you should pay attention to the following situations:

• Operate the explosive chemicals with great care and deal with them following the operating procedures.

• Try to control the rate of adding reactants and the temperature of reaction to avoid too vigorous reactions. Cooling measures may be taken if the reaction is difficult to be controlled.

• Check all the glassware carefully and replace the damaged or cracked one before assembling the reaction apparatus.

• Assemble the reaction apparatus correctly. The whole heating or reacting system could never be airtight under atmosphere⑫.

• On one hand, do not distill the liquid in the flask to be dry to avoid the explosion of glassware caused by local overheating⑬. On the other hand, peroxides or other explosive substances may form in the dry residue in the flask, which is also a risk of explosion.

• During the operation of vacuum distillation, do not use flat-bottomed flasks, Erlenmeyer flasks and other non-pressure-resistant glassware as the distillation or receiving flasks.

• Add zeolites to the flask prior to distillation. Do not add zeolites to a boiling liquid⑭.

(3) **Poisoning-proof**⑮

Most chemicals are at least slightly toxic, and many are very toxic and irritating if inhaled, ingested or contact with skin. Therefore, it is quite necessary to take safety measures against poison. To have a safe laboratory experience, you must know the following items to prevent poisoning.

① Inhalation

The experiment with poisonous chemicals and fumes, toxic vapors, or dust from finely powdered materials must be carried out in the hood or in a well-ventilated area. Make sure that the hood is working before the experiment starts.

② Ingestion

Never taste anything or pipet any liquid by mouth in the laboratory. No food or drink can be brought into the laboratory because they may be contaminated by the toxic chemicals.

③ Absorption through skin

Wear appropriate gloves and lab coat during the experiment. In case of unexpected skin contact with chemicals, wash the affected area with soap and running water thoroughly im-

---

⑫ 常压条件下，不能在密闭的体系中进行加热或反应，整个体系必须通大气。
⑬ 蒸馏时不要将蒸馏烧瓶中的液体蒸干，以免局部过热而引起玻璃仪器的爆裂。
⑭ 不要直接向沸腾的溶液中加沸石。
⑮ 防中毒。

mediately. Do not use organic solvents to rinse the chemicals on your skin[16]. Wash your hands with soap and water completely after you finish your experiment.

If you have symptoms of poisoning during the experiment, such as headache, dizziness and nausea etc. ,leave the laboratory and go to open area outdoor to get the fresh air. If the symptoms persist, seek medical help immediately.

(4) Prevent burns[17]

Skin may be burned by contact with hot, cold or corrosive substances, such as strong acid, strong base, and bromine. To avoid burns, it's best to wear appropriate gloves and protective glasses to prevent your skin and eyes from directly touching with chemicals. In case of such an accident, wash the burnt area immediately with copious amounts of running water, and then apply specific treatments as follows:

① Heat burns

Wash the burnt area with cold water for 10-15 min and then smear a burn ointment on it.

② Acid-injury

Wash the burnt area with 5% $NaHCO_3$ solution and then treat with a burn ointment.

③ Base-injury

Wash the burnt area with 1%-2% acetic acid or boric acid, wash with water again and then treat with a burn ointment.

④ Bromine-injury

Wash the burnt area immediately with alcohol, then smear with glycerol or a burn ointment.

⑤ Chemical splash in the eyes

Wear protective glasses at all times in the laboratory to avoid eye injury. If acid is splashed into your eyes, go to the eye wash station to wash your eyes immediately, then wash with 1% $NaHCO_3$ solution. For a base-injury, use water and then 1% boric acid for the eyewash.

(5) Prevent cuts and injuries[18]

An accident involving cut occurs when the glassware or apparatus is operated improperly. In case of cuts and injuries, deal with glassware by the following procedures:

① Break a glass rod or a glass tube correctly. Scratch a small line on one side of the rod or tube with a file and wet the line with water. Then, hold the rod or tube on both sides with a towel and quickly snap it by pulling the ends.

② Insert thermometer or glass tube into a stopper carefully and correctly. Wet the end of the thermometer or glass tube with water first, then hold the thermometer or the tube

---

[16] 不要用有机溶剂去冲洗粘在皮肤上的化学物质。
[17] 防灼伤。
[18] 防割伤。

with a towel close to the wet end and insert it into the stopper slowly with rotating firmly⑲.

③ Check the glassware carefully. This should be done before setting up the reaction apparatus. Glassware that is chipped on the rim should be discarded because one may be cut by the sharp edge.

④ Dispose of glassware properly. Discard the broken glassware and disposable pipets or capillaries to the properly labeled special container for broken glassware⑳.

In a case of cut injures, take out all the tiny pieces of glass first and rinse the wound with saline solution. Then, apply tincture of iodine to the wound and bind up with gauze. If the cut is large or deep, seek medical help immediately.

2. Disposal of Organic Laboratory Waste

After finishing an experiment, you may get a number of by-products in addition to the target product, such as filtrate from recrystallization, aqueous solutions from washing, recovered solvents, as well as used filter paper and drying agents, etc. It is an obligation and part of your lab work to properly dispose of the waste created by the experiments. The proper disposal of wastes not only avoids the occurrence of risk and minimizes the environmental impact, but also lowers the cost for handling the waste in organic laboratory. The handling of waste in the organic laboratory should be accomplished according to the following rules:

① The solid or liquid waste generated in the organic laboratory can be classified into many kinds. You should dispose of them in different ways. Never pour the waste down to the drain directly, such operation may cause a serious safety hazard㉑.

② There are various hazardous containers for disposing different kinds of waste in the organic laboratory. Different categories of liquid and solid waste should be poured into given labeled waste containers respectively㉒. Placing a waste in the wrong container not only leads to additional waste disposal costs but may also cause a dangerous reaction, resulting in safety and environmental hazards.

③ Organic solvents should be poured into properly labeled waste containers. Halogenated and nonhalogenated organic solvents must be placed in separate containers for recovery or disposal㉓. Halogenated waste container is only for disposal of organic wastes containing halogen atoms. The other solvents are usually placed in a flammable waste container.

④ Acidic or basic aqueous solutions from washing should be neutralized first to form a nonhazardous aqueous solution, then flush this solution down the drain with large volumes of water.

⑤ Solid waste contaminated with chemicals, such as used filter paper or drying agents

---

⑲ 安装温度计或玻璃管与塞子的连接装置时,应先用水将温度计或玻璃管的一端打湿,然后用毛巾在靠近湿润的一端握住温度计或玻璃管,慢慢旋转着将它们插入塞子。

⑳ 将破碎的玻璃仪器和一次性滴管或毛细管丢弃到盛放破碎玻璃器皿专用容器中。

㉑ 切勿将废物直接倒入下水道,这可能会造成严重的安全隐患。

㉒ 不同种类的液体和固体废物应分别倒入有指定标签的盛放废物的容器中。

㉓ 卤化的和非卤化的有机溶剂必须分开放置在不同的容器中,以便回收和处理。

should not be thrown in an ordinary trash can㉔. The exposed solid waste may bring a potential danger to personnel in the laboratory. Instead, hazardous solid waste should be thrown into designated and properly labeled trash can.

⑥ Some chemical waste has specific requirements for disposal. When in doubt about how to process those wastes, ask your instructor for help.

1.4 Pre-lab Reports, Notebooks and Lab Reports in the Laboratory

1. Pre-lab Reports㉕

It is important for you to prepare each experiment ahead of time. How well your performance in the laboratory depends largely on your preparation work. An adequate preparation can help you save time and effort during the experiment. The preparation work of each experiment includes reviewing the corresponding theory and principle, knowing the apparatus and operation of the experiments involved, seeking related information about the chemicals used in the experiment, and clarifying the experimental procedure. These tasks can be accomplished by completing a pre-lab report. A pre-lab report is not a simple copy of the contents described in each experiment of a textbook. By completing a qualified pre-lab report, you should be clear about what you are going to accomplish in the experiment, how you will do the experiment, and why you do each step㉖? In general, the experiment pre-lab reports include the following parts㉗:

① Heading: Provide information including the title of the experiment, your name and date.

② Objectives: State the purpose(s) of the experiments and list the experimental techniques as well as the major analytical approaches to be used in the experiment.

③ Principles: Write balanced chemical equations showing the overall process of the conversion of starting materials to products as well as by-products and further propose the reasonable mechanisms for these reactions.

④ Materials: Summarize the physical properties of reagents, solvents, product(s) as well as by-product(s), such as boiling points, melting points, density of a liquid compound, solubility of a solid compound in a certain solvent, etc. List the name and molar mass of each material, the weight(in grams) of each reactant and the volume of any liquid reactant, the molar amount of each reactant used. All of the above properties can be illustrated clearly in the fol-

---

㉔ 被化学药品污染的固体废物，如用过的滤纸或干燥剂，不能扔在普通垃圾桶内。
㉕ 预习报告。
㉖ 通过完成预习报告，学生应当弄清楚本次实验要做什么，怎样做，为什么要这样做。
㉗ 预习报告通常包括以下几部分：实验目的，实验原理，实验所用的试剂，实验流程图，实验装置图，产物的理论产量及预习题目等。

lowing Table 1.1.

Table 1.1 The properties of reactants, solvents, product(s) and by-product(s)

Materials	M_r	The amount of the materials			Physical properties			
		Volume used/mL	Weight used/g	Moles used	b.p./℃	m.p./℃	Density /(g/mL)	Solubility
A								
B								

⑤ Experimental procedure: Design a flow chart❷❽ that summarizes the sequence of operations involved with reaction processes and work-up procedures. A flow chart can help you to understand why each step in the procedure is performed and how the product is purified. You will know clearly at which step unwanted substances and unreacted starting materials are separated from the product(s) with the reaction flow chart.

⑥ Experimental apparatus: Provide a sketch of the assembly of glassware for the reaction and the corresponding operation.

⑦ Theoretical yield: The theoretical yield is the maximum possible amount of product that can be obtained assuming 100% conversion of the starting materials to product(s)❷❾. Theoretical yield of the desired product can be stated both in moles and in grams. Based on the number of moles used of the limiting reagent, the theoretical yield of the product(s) in moles can be obtained easily by the balanced equation of the reaction❸⓪. Then, the theoretical yield in grams will be calculated by timing the theoretical yield in moles with the molar mass of the product(s).

⑧ Pre-lab questions: Answer assigned pre-lab questions.

2. The Laboratory Notebook❸①

It is a vital part of your laboratory work to keep an accurate record of what you do and what you observe during the experiment. Experimental record is critical for you to understand and explain the results of the experiment later. Observations must be recorded in time while you are doing an experiment because it is very difficult to reconstruct the observed phenomena later. All of the observations and data for each step of the experiment should be recorded in a laboratory notebook. The following are general guidelines for a laboratory notebook.

① Start a new page of the notebook for each new experiment. Provide basic information including your name, the date, the title of the experiment at the top of each page.

② Write the balanced chemical equations of the reactions. Record the actual amounts of reagents used and calculate the corresponding theoretical yield accurately.

③ Provide a sketch of experimental apparatus for the experiment.

---

❷❽ 实验流程图。
❷❾ 理论产量是指假设起始原料100%转化为产物时,可能得到的产物的最大量。
❸⓪ 以用量相对少的原料为基准,根据反应方程式,可以很容易地得到以摩尔为单位的产物的理论产量。
❸① 实验记录本。

④ Record all experimental observations obtained during the experimental procedures, such as the appearance of reaction mixture(color, homogeneous or heterogeneous), reaction temperature and time, color changes, bubbling or smoke, precipitate formation, etc. Your observations can be recorded in tabular form with two columns, the left column is for procedure and the right one is for observations. Some important data, such as boiling points/melting points of the products, the appearance and quantities of the products should also be recorded in the right column in time.

⑤ It is unnecessary to copy detailed experimental procedures in your notebook from the textbook. You can write down the page number of the textbook on which the detailed procedures are elaborated and then only give an outline of the procedures that contains enough details. Note any variations from the textbook, such as the used quantities of the reagents, the modified experimental apparatus and procedures.

⑥ Make all records with a pen and do not delete anything you have written in the note using erase or white out. If you record the information incorrectly, cross it out and record the correct information on the side of it.

⑦ Record the observations and data directly on the notebook. Don't record the information on papers and then copy them into the notebook. The papers can easily be mixed up and the transcribing process is also prone to error.

3. Lab Reports and Samples of Lab Reports

(1) Lab reports

After you finish your work in the laboratory, a lab report is required to be completed. A lab report is not only a complete and accurate record of the experimental work that you do in the laboratory, but also a summary and interpretation of your experimental data and phenomena. A lab report can provide measurements to the success or failure of your experiment. Lab report generally includes the following parts[32]:

① Name and date of the experiment.

② Objectives: List the purpose and the techniques involved in the experiment.

③ Principles: Write down balanced chemical equations of the main reactions and important side reactions. Propose the mechanism of the reaction if applicable.

④ Materials: Summarize molecular weight, moles used, quantities used and physical properties(boiling point, melting point, density and solubility) of reagents, product(s), and by-product(s) in a tabular form.

⑤ Experimental flow chart: Construct a flow scheme of all operations in the experimental procedure.

⑥ Experimental apparatus: Sketch the assembly of glassware for the reaction and work-up operation.

---

[32] 实验报告通常包括以下几部分：实验名称，实验目的，实验原理，所用的试剂，实验流程图，实验装置，理论产量，实验步骤与现象，结果与讨论及思考题等。

⑦ Theoretical yield.

⑧ Procedures and observations: List the detailed steps of the experiment and keep a record of what you have observed from the reaction. This record may be somewhat different from the related parts in the pre-lab report. You should record exactly what you have done and what you have seen during the experimental process.

⑨ Results and discussion: Summarize the quantities and physical properties of the product isolated in the experiment, which includes boiling point and/or melting point, odor, color as well as crystalline form, if the product is a solid. The percent yield should also be calculated for a synthesis reaction. The percent yield is a direct and important measure to evaluate a chemical synthesis. It is defined as the ratio of the actual obtained mass of the isolated product to its theoretical yield❸.

$$\text{Precent yield} = \frac{m_{\text{obtained}}}{m_{\text{theoretical}}} \times 100\%$$

The experimental results should be discussed from the following aspects❹: (a) analysize your data and the phenomena that are different from those described in the textbook; (b) list the problems and the corresponding solutions in the experiment; (c) comment on the source of error in your experiment and summarize the problems that should be noticed in the next experiment; (d) give some suggestions, if applicable, for improvement of the experimental design.

⑩ Post-lab assignments.

(2) A sample of lab report

Preparation of *n*-Butylbromide   Name: Date:

Objectives:

1. To learn the principle of the preparation of *n*-butylbromide.
2. To practice the assembling of refluxing apparatus with a gas-trap.
3. To learn the principle and operation of the extraction of liquid compounds.
4. To practice the fundamental operations of distillation, washing and drying.

Principle:

The common method for converting alcohols into alkyl halides is to treat the alcohols with HX(X=Cl, Br, or I), a typical nucleophilic substitution reaction. This reaction with primary alcohols works best via the S_N2 pathway and heating must be used for the reaction to be completed. In this experiment, *n*-butyl alcohol will be converted to *n*-butyl bromide by reaction with sulfuric acid and sodium bromide. The main reactions are shown as below:

---

❸ 产物的实际产量与理论产量的百分比称为产率。

❹ 实验的结果与讨论包括以下几个方面：(a) 对实验结果和数据进行分析；(b) 实验中出现的问题和解决问题的方法；(c) 实验中所见错误原因的讨论和下次实验需要注意的问题的总结；(d) 对实验提出建设性的意见。

$$NaBr + H_2SO_4 \longrightarrow NaHSO_4 + HBr$$

$$HBr + n\text{-}C_4H_9OH \xrightleftharpoons{\triangle} n\text{-}C_4H_9Br + H_2O$$

This reaction is reversible. A large excess of acid is normally used to drive the equilibrium to the right. Accompanying with the main reactions, some side-reactions may readily occur.

$$2HBr + H_2SO_4 \xrightarrow{\triangle} Br_2\uparrow + SO_2\uparrow + 2H_2O$$

$$CH_3CH_2CH_2CH_2OH \xrightarrow[\triangle]{H_2SO_4} CH_3CH_2CH=CH_2 + CH_3CH=CHCH_3$$

$$2CH_3CH_2CH_2CH_2OH \xrightarrow[\triangle]{H_2SO_4} (CH_3CH_2CH_2CH_2)_2O + H_2O$$

These side reactions can be minimized by controlling the reaction temperature and the concentration of sulfuric acid used. Moreover, the gas-trap system also needs to be equipped to the reflux apparatus to absorb the hazardous gas generated during the reaction.

Materials:

Table 1 The dosage of reagents and product

Materials	M_r	Volume used/mL	Weight used/g	Mole used/mol
n-Butyl alcohol	74.12	6.2	5.0	0.068
Sodium bromide	102.93		8.3	0.08
n-Butyl bromide	137.03			

Table 2 The physical properties of reagents and product

Materials	b.p./℃	m.p./℃	Density /(g/mL)	Solubility		
				Water	Ethanol	Ethyl ether
n-butyl alcohol	117.3	−89.5	0.8098	9^{15}	∞	∞
Sodium bromide		747.0		116.0^{50}	slightly soluble	×
n-Butyl bromide	101.6	−112.4	1.2758	0.06^{16}	∞	∞

Experimental Apparatus:

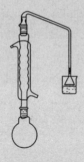

Reflux and gas-trap apparatus

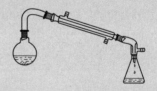

Simple distillation apparatus

Experimental Flow Chart:

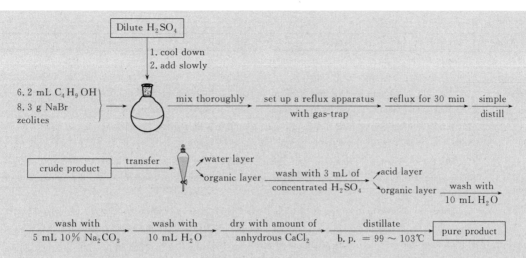

Theoretical Yield:

$$n\text{-}C_4H_9OH + NaBr + H_2SO_4 \xrightarrow{\Delta} n\text{-}C_4H_9Br + NaHSO_4 + H_2O$$

M_r 74.1 M_r 102.9 M_r 137.0

0.068 mol   0.08 mol

moles of $n\text{-}C_4H_9Br = 0.068$ mol

theoretical yield $= 0.068$ mol $\times 137$ g/mol $= 9.32$ g of n-Butyl bromide

Procedures and Observations:

Time	Procedures	Phenomena
8:30	1. Place 10 mL of water in a beaker, add 10 mL concentrated sulfuric acid slowly and cool the solution to room temperature.	Heat is released.
	2. Place 8.3 g of sodium bromide in a 100 mL round-bottom flask, add 6.2 mL of n-butyl alcohol. Dilute sulfuric acid is added slowly while shaking. Add several zeolites to the mixture.	The mixed solution is clear. NaBr can not be dissolved completely
	3. Set up a reflux and gas-trap apparatus. Place some water in the beaker	
9:00	4. Heat the mixture until the mixture begins to reflux gently. Heat the solution to reflux for 30 min. The flask should be shaken frequently during this period.	The mixture boils slightly and NaBr dissolves slowly. The liquid in the flask is separated into three layers.
9:34	5. Turn off the heating mantle and cool down the mixture. Then remove the condenser with the gas-trap.	There are two layers in the flask, the upper layer is orange-yellow and the bottom layer is clear.
9:40	6. Assemble a simple distillation apparatus. Add several zeolites to the flask again.	The upper layer of liquid in the flask disappears gradually. Turbid distillate is collected in the receiving container.
	7. Distill the mixture until the distillate appears to be clear.	
9:55	8. Transfer the distillate to a separatory funnel, add 10-15 mL of water to the mixture and allow the layers to separate.	The organic layer (bottom) is turbid.

Time	Procedures	Phenomena
	9. Separate the mixture and save the organic layer (bottom) in another dry separatory funnel.	
	10. Wash the saved organic layer with 3 mL of concentrated sulfuric acid and remove the acid layer (bottom).	The organic layer (top) is pale yellow and the acid layer (bottom) is red-brown.
	11. Wash the organic layer with 10 mL of water again, then with 5 mL of 10% Na_2CO_3 and finally with 10 mL of water.	The organic layer is turbid and always is on the top.
	12. Dry the crude n-butyl bromide using a small amount of anhydrous calcium chloride in a dry Erlenmeyer flask until the liquid is clear.	The crude product is cloudy and then becomes clear after being placed for a period of time for drying.
11:10	13. Decant the liquid into a dry distillation flask. Distill and collect the fraction that boils between 99-103 ℃.	Clear and transparent distillate is obtained with the boiling point at a stable reading of 100-104 ℃.
11:30	14. Measure the volume of the product, and calculate the percentage yield.	$V=4.1$ mL, $m=5.2$ g

Results and Discussion:

1. Appearance of the n-butyl bromide: clear and transparent liquid.

2. Actual yield of n-butyl bromide: $m=4.2$ g.

3. Boiling points of n-butyl bromide: 100-104 ℃.

4. Percent yield of n-butyl bromide:

$$\text{Precent yield} = \frac{5.2}{9.3} \times 100\% = 56\%$$

5. Discussion

① With the conversion of n-butyl alcohol to n-butyl bromide, three layers appear in the reaction flask. The top layer is the formed n-butyl bromide, the middle layer may be the n-butyl hydrogen sulfate produced by the reaction of n-butyl alcohol with sulfuric acid. With the conversion of n-butyl alcohol into n-butyl bromide, the middle layer will disappear gradually.

② At the end of the reaction, both the upper and bottom layers of liquids in the reaction flask turn orange-yellow, which is caused by the bromine produced by the side reaction of hydrogen bromide with concentrated sulfuric acid.

③ In the work-up procedure, concentrated sulfuric acid is used to remove the unreacted n-butyl alcohol and the by-products of 1-butene, 2-butene and dibutyl ether from the crude n-butyl bromide. This step should be carried out in a dry separation funnel.

④ The crude n-butyl bromide must be dried completely prior to the final distillation to avoid the formation of azeotrope between water and n-butyl alcohol, which will affect the purity of the final product.

⑤ There are two reasons why anhydrous calcium chloride is used as a drying agent. Firstly, it not only has high water absorbing capacity, but also is very cheap. In addition, anhydrous calcium chloride can form a complex with the trace of n-butanol alcohol that has not been removed during the washing process, which also contributes to further purification.

⑥ It is recommended to keep all the layers obtained from different washing steps. Never discard any layer until the experiment has been completed.

1.5 Common Laboratory Glassware and Equipment

1. Laboratory Glassware

Organic chemistry experiments are commonly conducted with a variety of glassware. You will find many kinds of glassware when you walk into the laboratory. Among these, the most commonly used glassware for organic chemistry experiments is standard-taper glassware with ground-glass joint[35]. The joints of the standard-taper glassware include an inner joint and an outer joint. The former one has an outward side covered with ground glass surface, and the latter one has an inner side covered with ground glass surface. The standard-taper glassware is classified by the symbol ℻ followed by two sets of numbers separated by a slash, such as ℻ 19/22. Here, the first number after the symbol ℻ represents the diameter of the joint in millimeters at its widest point, and the second number is the length of the ground glass surfaces in millimeters[36]. So, ℻ 19/22 means that the joint has a maximum inside diameter of 19 mm and the length of 22 mm.

Standard-taper glassware is available in a variety of sizes. All the joints in standard-taper glassware are ground so that all the pieces of the same size can exactly fit together. Some standard-taper glassware commonly found in the organic laboratory is listed in Figure 1.1.

Among these standard-taper glassware[37]:

① Round-bottom flask is usually used for distillation or reflux. For rather complicated reactions where multiple operations needed to be performed simultaneously, round-bottom flask with three necks[38] is preferred.

② Erlenmeyer flask is used as a common receiver of simple distillation. It is also often used in some work-up procedures, such as recrystallization, washing and extraction, etc.

③ Thermometer adapter is used to hold a thermometer and joint it with other standard-taper glassware.

④ Distilling head is used for simple distillation. It has three joints. The upper joint is for holding a thermometer, and the other two joints are used to connect the distillation flask to the condenser so as to make the distillate pass into the condenser.

⑤ Adapter is commonly used to connect the condenser to a receiver in different types of distillation. Claisen adapter can also be used to place two pieces into one joint of a reaction

---

[35] 标准磨口玻璃仪器。
[36] 符号后面的第一个数字代表磨口最大端的直径（毫米），符号后面的第二个数字代表磨口的长度（毫米）。
[37] 常用的标准磨口玻璃仪器有：圆底烧瓶，锥形瓶，温度计套管，蒸馏头，接引管，冷凝管，滴液漏斗，分液漏斗，分水器及色谱柱等。
[38] 圆底三口烧瓶。

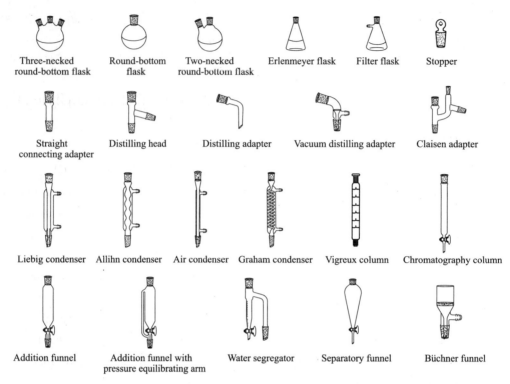

Figure 1.1　Some common standard-taper glassware in the organic laboratory

flask at the same time.

⑥ Condenser is required to condense vapors to liquids. Liebig condenser[39] and air condenser are both for distillation. Allihn condenser[40] is used for reflux during a reaction.

⑦ Addition funnel can be used to add liquid chemicals to the reaction system.

⑧ Separatory funnel is for some reaction work-up, such as extractions and washing.

⑨ Water segregator can remove the water formed during the reaction from the reaction system while the reaction is going on.

⑩ Büchner funnel is usually used to separate solids from liquids during the operation of vacuum filtration[41].

⑪ Filter flask is a container to collect the filtrate during vacuum filtration.

⑫ Chromatographic column is for column chromatography separation.

2. Common Glassware in the Organic Laboratory

In addition to the standard-taper glassware introduced above, there are also some common glassware that are used widely in the organic laboratory. Some common glassware mostly found in the organic laboratory is listed in Figure 1.2.

All experiments carried out in the organic laboratory require certain items of equipment

---

[39]　直形冷凝管。

[40]　球形冷凝管。

[41]　减压过滤。

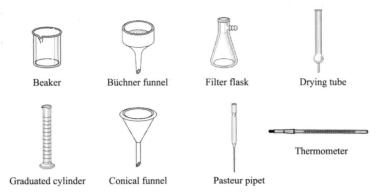

Figure 1.2 Some common glassware in the organic laboratory

and glassware. It is crucial to assemble these various glassware and equipment together correctly and tightly according to the experimental apparatus. Various clamps(Figure 1.3) and iron support[42] are needed to hold these items during the reaction process in the organic laboratory. Clamps that hold different pieces of glassware can be attached to the rod of the iron support. The adjustable ring-angle clamp holders[43] are mostly used to attach the clamps to the iron support. Different kinds of clamps are used for holding different kinds of labware. For example:

① Three-finger clamps[44] are used to hold the various reactors, such as round-bottom flasks, three-necked round-bottom flasks or Erlenmeyer flasks.

② Ring clamps[45] can be used to hold separatory funnels and regular funnels.

③ Plastic keck clamps are used as needed to secure the union of the glassware joints firmly.

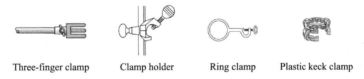

Figure 1.3 Some clamps and clamp holders in the organic laboratory

The assembly of the apparatus should begin with the reaction or distillation vessel first. It is recommended to assemble the whole apparatus with one iron support so that the whole apparatus can be moved easily if needed. Adjust the clamps to hold the labware firmly. Do not overtighten the clamps to make sure that the glassware and equipment can be adjusted freely and safely.

---

⑫　铁架台。

⑬　双口夹（双顶丝）。

⑭　三爪夹（烧瓶夹）。

⑮　铁圈。

3. Cleaning of Glassware

Upon the completion of a reaction, you should dispose of all chemical residues properly and then wash the glassware used in the experiment timely. Most chemical residues can be removed from the glassware by scrubbing the glassware with scouring powder and hot water. Acetone is also often used in the organic laboratory to remove the residues which are difficult to be washed off by water. When you wash the glassware with acetone, wear gloves and dispose of the washed acetone in the proper waste container. Acetone should be used as little as possible to avoid waste and environmental pollution.

The solution of alcoholic sodium hydroxide[46] is a more powerful and effective cleanser to remove grease and stubborn organic residues from glassware in the organic laboratory. This solution is prepared by adding some solid sodium hydroxide to a vessel containing ethanol. The glassware needs to be washed and immersed in the cleaning solution for several hours, and then rinsed thoroughly with water to complete the process. The solution of alcoholic sodium hydroxide is strongly basic, wear gloves and eye protection to avoid severe burns while handling it.

4. Drying of Glassware

Some organic reactions are sensitive to water and must be carried out with dried glassware. The simplest way to ensure glassware dry is to leave the clean glassware in the air to dry naturally. This method is convenient but time-consuming. Wet glassware can be dried more quickly by being heated in an oven at about 100 ℃ for some time. Be noticed that the glassware in the oven is very hot, so remember to wear asbestos gloves[47] when taking the dried glassware from the oven. In addition, allow the dried hot glassware to cool down to room temperature before using it for a reaction.

When there is an urgent need for dried glassware in the laboratory, you can dry the wet glassware as follows: rinse the glassware with a small amount of low boiling point solvent which can be miscible with water[48], such as alcohol or acetone. Then, pour out the solvent and blow the glassware with an electric blower. The solvents used for rinsing the glassware can be recovered and applied to wash the organic residues later[49].

1.6 The Literature of Organic Chemistry

Chemical literature is the summary and record of scientific research practice in chemistry. The literature is built upon past work and is a valuable resource. Literature review is a key element when a chemist attempts to develop a chemical research project. By consulting

---

[46] 氢氧化钠的乙醇溶液。
[47] 石棉手套。
[48] 和水互溶的低沸点溶剂。
[49] 用于润洗玻璃器皿的溶剂可以回收,后续再用于清洗有机残留物。

the relevant literature, a chemist would know if the project to be carried out has been done previously by others. If it has been made, he or she needs to further study the literature related to the project to master the development process and the latest research findings. Analyzing and summarizing the reported research methods not only broadens our horizon, but also helps avoid doing unnecessary repetitive research work. It is essential for chemists to find the information we need from the large amount of existing literature.

As a beginner in organic laboratory, it is also necessary for you to learn about the most commonly used reference books and literature in organic chemistry and master the consulting methods. Before an experiment, you can refer to the pre-lab resources for information like physical properties, safety and disposal method for the chemicals, which will help you to better understand the experimental procedure and safety requirements. After the experiment, you can also find alternative methods for preparing the desired product(s) that you have obtained in the experiment from related work by others, then inspect your experimental results from different perspectives. Compare these various methods critically in terms of availability of reaction reagents, complexity of equipment, ease of workup, and safety issues, etc., thus to improve your ability to design and carry out comprehensive experiments.

There are three types of literature sources including handbooks, chemistry journals, and chemical databases. With the rapid development of digital techniques, the way we access information has revolutionized. One can also visit online sources to find the desired literature rapidly and comprehensively.

1. Reference Books

Handbooks are the traditional sources of information on chemicals. Some handbooks are particularly useful for organic chemistry experiment:

① Aldrich Catalog Handbook of Fine Chemicals, published biennially by the Aldrich chemical company. It contains thousands of organic and inorganic compounds with their physical properties, safety and disposal information, and the references to IR, UV, NMR spectra.

② John R. Rumble, CRC Handbook of Chemistry and Physics, 100th ed., published annually by CRC press: Boca Raton, FL. 2019. It contains more than 12,000 organic compounds with physical properties, solubilities and structural formulas.

③ M. J. O'Neill, A. Smith, P. E. Heckelman, J. R. Jr. Oberchain, (Eds.), The Merck Index: An Encyclopedia of Chemicals, Drugs, and Biologicals, 14th ed., 2006, published by Merck & Co., Inc.: Whitehouse, NJ. It gives physical properties and references to syntheses, safety information, and application of more than 10,000 organic compounds.

④ Organic Syntheses, published by Wiley: New York, 1932—present. It is a compilation procedure for preparing specific compound or one similar to it. The especially useful details of each procedure are written in footnotes at the end.

⑤ T.-L. Ho, Fieser's Reagents for Organic Synthesis, 24 vols., 1967—2008, published by Wiley: New York. It provides improved information of the preparation and purification of organic compounds. Many newer reagents are covered in this series for safer and easier reaction processes in the preparation of traditional organic compounds.

2. Chemistry Journals

Important current journals published in organic chemistry mainly include the following:
① Chinese Journal of Organic Chemistry
② European Journal of Organic Chemistry
③ Journal of the American Chemical Society
④ Journal of Heterocyclic Chemistry
⑤ Journal of Medicinal Chemistry
⑥ Journal of Organic Chemistry
⑦ Organic & Biomolecular Chemistry
⑧ Organic Letters
⑨ Synthesis
⑩ Synthetic Communications
⑪ Tetrahedron
⑫ Tetrahedron Letters

All these journals are available online. It also contains supplemental information that provides electronic access to detailed experimental procedures and important spectral data.

3. Chemical Databases

There is a large amount of literature in organic chemistry, which makes it difficult to find desired information among the tremendous literature. Chemical databases provide an invaluable way to get all the information about specific chemical compound by looking up journal articles of the entire literature.

The Chemical Abstracts(CA), annually published by the American Chemical Society, provides the most comprehensive abstracts of original research work from journals, books, conferences and patents. The desired information of a chemical can be indexed in Chemical Abstracts by registry number, chemical substance, molecular formula, author's name, research topic or patent numbers.

Chemical Abstract Services(CAS) also provides a number of databases for Chemical Abstracts. The mostly used of these databases is SciFinder Scholar(scifinder.cas.org), which is an excellent and invaluable search engine. You will have the most efficient way to survey Chemical Abstracts and other chemistry journal literature for organic compounds by SciFinder Scholar. Besides SciFinder Scholar, STN is another very helpful database provided by CAS.

4. Online Resources

There are many internet websites where various kinds of information including online libraries, online periodical resources, online patent resources and database resources about organic compounds can be readily accessed, such as http://www.lib.pku.edu.cn/html, http://www.chinajournal.net.cn/, http://www.acs.org/, http://www.chemsoc.org/, and http://www.chemistry.rsc.org.rcs/, etc. You can find a list of reviewed web sites on the online site of the Journal of Chemical Education(www.jce.divched.org.).

1.7 Green Chemistry in the Organic Laboratory

Green chemistry is the set of fundamental principles aimed to reduce or eliminate hazardous substances in the design, manufacture, and application of chemical products[50]. The ultimate goal of green chemistry is to lead environmentally friendly chemical processes. Green chemistry is based on twelve principles that were first introduced twenty years ago. The twelve principles include: prevention, atom economy, less hazardous chemical synthesis, designing safer chemicals, safer solvents and auxiliaries, design for energy efficiency, use of renewable feedstocks, reducing derivatives, catalysis, design for degradation, real-time analysis for pollution prevention, and inherently safer chemistry for accident prevention[51].

It is clear that the sustainable development of chemistry is largely depended on the advances in green chemistry. It is essential for undergraduate students majoring in chemistry to have awareness of green chemistry. They should not only be able to accomplish an experiment successfully in the laboratory, but also have the abilities to apply the principles, necessary knowledge and skills in green chemistry to direct their laboratory work. In order to cultivate students' awareness of green chemistry and help them to face the environmental challenges ahead, the education in green chemistry should be expanded upon the organic chemistry experiment at the undergraduate level from the following aspects.

1. Less Hazardous Organic Synthesis

In order to reduce or eliminate the hazards from chemical reaction processes to human health as well as environment, synthetic methods, if possible, should be designed to use and generate substances with less toxicity. In recent years, many new reactions and experimental techniques have been developed to replace the traditional ones in organic laboratory to meet the requirements of green chemistry.

(1) Replacement of solvents

The greener replacement of solvents may be one of the most active areas of research in green organic chemistry. We can predicate whether a substitution solvent is greener and more suitable from these aspects, such as the safety information (safety and toxicity), physical

---

[50] Anastas P T, Warner J C. Green Chemistry: Theory and Practice [M]. New York: Oxford University Press, 1998.

[51] 绿色化学的十二条原则：防止污染，原子经济性，减少有毒有害物质的使用，设计安全的化学产品，使用安全的溶剂及辅助物质，反应能耗低，采用可再生的原料，减少副产品，使用催化剂，产品可降解，实时分析预防污染，降低化学过程中潜在的危险。

properties and the relative costs of waste solvent disposal. There is no doubt that water is the ideal alternative solvent because of its available and nonhazardous. It has been reported that many organic reactions are able to be performed in aqueous solutions now. As we all know, the major inconvenience of using water as solvent for the organic reaction is that most organic reactants are insoluble or only slightly soluble in water, which leads to a heterogeneous reaction with some disadvantages, such as slow reaction speed and low conversion rate, etc. However, this problem can be solved well by stirring the reaction mixture vigorously or adding a phase transfer catalyst to the reaction system.

Besides water, other eco-friendly solvents, such as supercritical carbon dioxide and various ionic liquids are also used as ideal alternatives for traditional organic solvents. Solvent-free reactions, which can eliminate the need for any solvents, are now the focuses and forefront projects in academic research of green organic chemistry. The examples of greener replacement of solvents are described in some experiments in Chapter 5 of this book.

(2) Less hazardous materials

Many innovative methods have been studied in undergraduate laboratory to synthesize organic compounds using less hazardous materials. A good example is optimizing the synthetic pathway towards hexanedioic acid. Hexanedioic acid is usually prepared by the oxidation of cyclohexene with various oxidizers. The most common oxidizing agents used in the laboratory are sodium dichromate and potassium permanganate. These two oxidizers are both derived from heavy metals, which are commonly toxic and environmentally hazardous. The disposal of the oxidizers and their reduction products is therefore troublesome and expensive. Fortunately, a method has been reported using hydrogen peroxide as a green alternative oxidizing reagent for sodium dichromate or potassium permanganate in the preparation of hexanedioic acid(Scheme 1)[52].

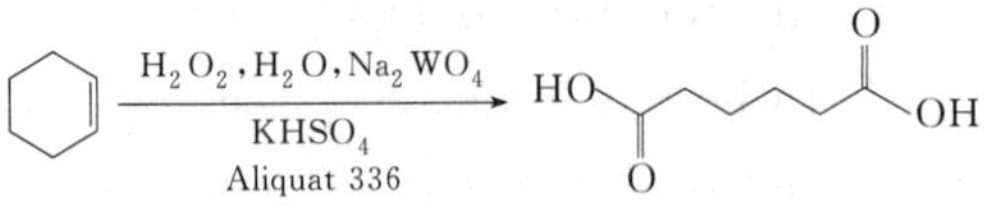

$$\text{cyclohexene} \xrightarrow[\text{KHSO}_4 \\ \text{Aliquat 336}]{H_2O_2, H_2O, Na_2WO_4} HOOC(CH_2)_4COOH$$

(Scheme 1)

This replacement reduces the hazards of heavy metals to human health and environment. At the same time, it also helps save the cost of waste disposal. Although a phase transfer catalyst is needed here to enhance the reaction rate of this heterogeneous reaction, the catalyst can be recovered directly from the aqueous solution and be recycled several times.

2. Reducing Energy Requirements

Energy requirement of a chemical process should be minimized considering its impacts

---

[52] Scott M R, James E H. Green Chemistry in the Organic Teaching Laboratory: An Environmentally Benign Synthesis of Adipic [J]. J. Chem. Educ., 2000, 77 (12): 1627-1629.

on environment and economy. Reactions that do not require a great amount of energy are highly desirable and environmentally friendly. It is best to carry out a reaction at ambient temperature and atmospheric pressure with no need for extra energy input. To reduce energy consumption of chemical synthesis, some reactions with new energy input methods, such as microwave-assisted and ultrasonic-assisted reactions㊃, electroorganic, and photochemical organic synthesis㊄ have been developed in organic laboratory. Some examples of these reactions are described in detail in chapter five of this book.

3. Recycling of Experimental Products㊅

It has always been a headache in the organic laboratory to deal with the experimental products handed in by students after each experiment. The accumulation and long-term storage of these chemical products is not only a waste of resources, but also poses potential threats to the environment and safety of the laboratory. How to recycle the experimental products effectively is an urgent problem for organic laboratory. One of the effective ways of doing this is to design a series of comprehensive experiments by combination of several isolated synthesis experiments, using the product(s) of one reaction as the reactant for another one. An example is illustrated below(Scheme 2).

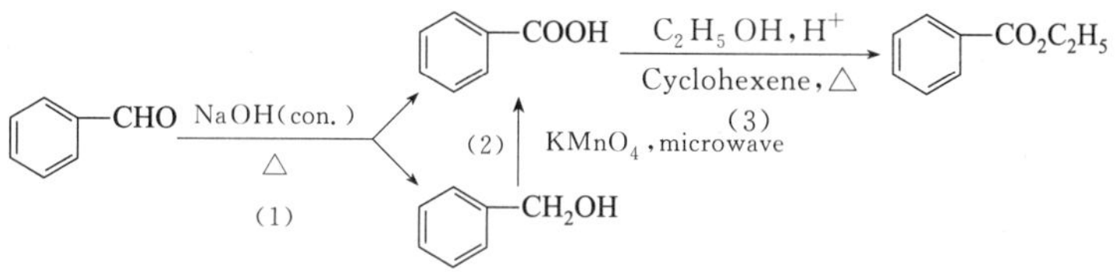

(Scheme 2)

The experiment series include three steps. The first step is a Cannizzaro reaction of benzaldehyde to form the products of benzoic acid and benzyl alcohol. The second step is the conversion of obtained benzyl alcohol to benzoic acid by oxidation reaction. In the last step, ethyl benzoate is prepared by the esterification㊆ of benzoic acid formed in the second step with ethanol using a catalytic amount of concentrated acid. Such a design reduces the risk of storing large quantities of chemicals in the laboratory and costs of handling them, following the requirements of setting up different types of reactions(Cannizzaro reaction, oxidation and esterification reaction)in the series of experiments.

4. Miniscale Experiments㊇

Miniscale experiments have become popular in undergraduate organic laboratory recently. Miniscale experiments are those kinds of experiments which generally use approxi-

---

㊃ 微波和超声波辅助的反应。
㊄ 有机电化学和有机光化学反应。
㊅ 实验产品的再利用。
㊆ 酯化反应。
㊇ 微型实验。

mately 0.3-5 g of starting materials and are performed with the scaled-down macroscale glassware. Comparing with conventional experiments, miniscale experiments are greener ones due to the following advantages: ① Miniscale experiments are less expensive and environmentally friendly. Miniscale experiments using minimal amounts of staring material usually require fewer reagents and less solvent for the reactions and work-up procedures, as a result, relatively little waste is generated from the whole processes. ② Miniscale experiments are safer. The smaller quantities of chemicals used in the miniscale reactions reduce safety risks and contamination caused by hazardous chemicals. ③ Miniscale experiments are also generally more energy-efficient. Miniscale experiments, when conducted properly, will need less time and energy for the reaction to complete. Also, the subsequent work-up and purification procedure will be simplified as well.

Reactions with innovative methods of green chemistry are now undergoing a rapid development worldwide. In addition to what we have discussed above, there are some other important principles and techniques, such as atom economy, catalysis and environmental impact factor, etc. are closely related to green organic chemistry. You will have greater opportunities to learn and apply these knowledge and principles in your future laboratory work.

Chapter 2
Basic Operations of Organic Chemistry Experiments

2.1 Heating and Cooling

Heating and cooling are important techniques extensively adopted in organic laboratory. For instance, most organic reactions cannot occur at room temperature so heating is required to reach a desired temperature. Whereas, some reactions must take place at a low temperature to avoid undesired side reactions and to make exothermic reactions be controllable. Moreover, heating and cooling methods are also utilized during the work-up procedure of reactions, such as distillation, evaporation and recrystallization, etc. Several frequently used heating and cooling methods are summarized in the following part.

1. Heating

Most organic substances, especially solvents, are highly flammable. In this case, it is restricted to heat organic compounds with open flames in organic laboratory. The following heating methods are the safest ways to heat organic compounds, all of which involve no open flame, and can therefore reduce the risk of fire hazards in organic laboratory[58].

(1) Hot plates

Hot plates can be used as a direct heat source for the heating of flat-bottomed containers. Stirrer hot plates, i. e., hot plates co-assembled with magnetic stirrers, are usually used in organic laboratory to achieve stirring and heating at the same time.

(2) Heating baths

There are several kinds of heating baths in the laboratory, including water bath, oil bath and sand bath, etc.

① Water baths

Water bath may be used when the required temperature is not higher than 90 ℃. The flask to be heated should be submerged in a water bath with the surface of the liquid in the flask slightly

---

[58] 有机化学实验室不能使用明火加热,以降低火灾风险。

lower than the water surface. The temperature of the water bath is usually 10 ℃ higher than that of the mixture in the flask[59].

During the heating process, water in the container will evaporate rapidly, making it necessarily important to check the water level frequently and add water in time to avoid safety risk. Aluminum foil is usually used to cover the open portions of a water bath and minimize the loss of water.

② Oil baths

Another kind of liquid bath is oil bath. Oil is inert and nonvolatile, which makes oil bath more widely used than water bath in laboratory. Meanwhile, oil bath can be combined with stirring hot plates, which makes simultaneous heating and stirring of a reaction mixture possible[60].

Both mineral oil and silicon oil are commonly employed for oil bath. Silicone oil is stable and can be heated to 200-275 ℃ without reaching its flash point, and is therefore generally ideal for oil bath. Although mineral oil is cheaper than silicone oil, it is flammable and cannot be heated above 175 ℃, thereby terribly limiting its application in organic laboratory. When an oil bath is employed for heating, be careful not to let hot oil connect with water to avoid hot oil spattering[61].

Silicon oil and mineral oil are both water-insoluble, making it troublesome to remove the oil film from the outer surface of the flask to be heated. The flask can be wiped by using a small amount of dichloromethane on a paper towel, and then washed with soap and water.

③ Sand baths

Sand bath is prepared by putting about 1-3 cm of sand in a flat-bottom container and then placing the container on a hot plate. A temperature gradient exists along the various depths of the sand due to its poor heat conductivity. The temperature is generally monitored with a thermometer inserted into the sand bath at the same depth as the flask being heated[62]. Given that the heated glass containers are prone to burst at high temperature, sand baths should not be heated above 300 ℃. In addition, stirring hot plates cannot be used in sand bath.

(3) Heating mantle

Heating mantle is a common device for heating round-bottom flasks, typically in a half-sphere shape with a variety of sizes to fit different sizes of round-bottom flasks. The heat is provided by electric wires embedded within a layer of fabric. The heating rate can be controlled by adjusting a variable transformer connected with the heating mantle. As a powerful and safe heating method, heating mantle can provide the heated flask with uniform heating.

---

[59] 将要加热的烧瓶浸在水浴中，且烧瓶内液体的液面应略低于水面。水浴温度通常比烧瓶内混合物的温度高10 ℃。

[60] 油浴可以和电磁搅拌器联合使用，从而使反应过程中的加热和搅拌操作可以同时进行。

[61] 当使用油浴加热时，要防止水或有机溶剂滴入油锅，以免热油飞溅。

[62] 由于沙子的导热性较差，所以沿沙浴的不同深度存在温度梯度。用温度计测温时，要将温度计插入沙浴中，并保持深度与被加热的烧瓶相同。

Meanwhile, the risk of glassware shattering can also be reduced due to the indirect contact of the heated flask with the electric wires❻❸.

Heating mantle has a high heat capacity, so it is time-consuming to cool the hot flask only by turning off the heating mantle. The heating mantle should be removed immediately from below the flask after being turned off, so as to allow the hot flask to cool quickly in the air or by means of cooling bath. It is absolutely forbidden to transfer organic substances above the hot heating mantle to avoid short circuit or leakage due to the spilling of substances on the layer of fabric❻❹.

2. Cooling

Cooling is always required to control exothermic reactions, or to lower the temperature of hot reaction mixtures so as to perform the following work-up procedure in time. Cooling is also employed to maximize the recovery of crystalline solid products from a recrystallization operation.

Cooling baths are frequently used methods in organic laboratory. Water bath is the common one to cool a mixture to room temperature, while ice-water bath, which can be simply prepared by adding crushed ice to a water bath, is rather effective for cooling and can be used to achieve the temperature of 0 ℃. It is noted that ice cannot contact completely with the vessel being cooled, and is thus forbidden to be used alone as coolant.

A better cooling effect with temperatures from −20 ℃ to 0 ℃ can be achieved by using ice-salt bath, which is usually prepared by mixing sodium chloride into an ice-water bath❻❺. Other salts, such as calcium chloride, and ammonium chloride can also be combined with ice to achieve different lower temperatures. The proportion ratio of the salts to ice will affect the cooling temperature. The lowest temperatures abstained by the mixture of ice with other different salts are shown in Table 2.1.

Table 2.1 The lowest temperature of different ice-salt mixtures

Salts	$m_{salts} : m_{ice}$	Lowest temperature/℃
NaCl	1 : 3	−40
NH_4Cl	1 : 4	−15
$CaCl_2 \cdot 6H_2O$	1 : 0.7	−55
$NaNO_3$	1 : 2	−18
NH_4NO_3	1 : 2	−17

Even lower temperatures can be achieved by mixing dry ice or liquid nitrogen with some organic solvents. For instance, dry ice mixed with acetone can reach −78 ℃, and the lowest temperature

---

❻❸ 作为一种高效、安全的加热方式，电热套具有仪器简单、操作方便、加热迅速且基本均匀等优点。同时，电加热套在使用过程中，被加热的反应瓶不直接与电线接触，这也降低了玻璃器皿在加热过程中碎裂的风险。

❻❹ 不要在电热套上方转移试剂，以免药品洒在电热套内引起短路或漏电。

❻❺ 如果体系需要更低的温度（−20 ℃），可以采用冰盐浴冷却。冰盐浴的制法通常是将粉碎的冰和相当于其三分之一质量的粗食盐混合。

of the cooling mixture of liquid nitrogen and acetone can be as cold as −196 ℃[66]. These special cooling baths should be contained in a Dewar flask, i. e. , a double-walled vacuum chamber. Dry ice or liquid nitrogen should be slowly and carefully added to organic solvents in the Dewar flask while stirring to obtain a uniform mixture. Be sure to wear protective glasses and asbestos gloves during this process to avoid frostbite.

2.2 Drying Organic Solutions and Solids

1. Drying Solids

Most of the solid organic products are finally isolated from the mixtures using some separation techniques, such as recrystallization, filtration, sublimation and chromatographic separation, etc. It is rather difficult to get absolutely dry products through the above operations. The residual water and organic solvent in the final products may affect their weight, melting point, quantitative elemental analysis and spectra[67]. In this case, the obtained solid products must be further dried. The methods used to dry organic solids depend on the properties of the solids, the nature of the residual solvents in the solids, and the availability of instrument in the laboratory. The following methods are commonly used in the organic laboratory to dry solid compounds.

(1) **Air dryness**

Air dryness means that a solid is dried in the air by simply spreading the solid on a piece of filter paper or on a clean watch glass. The solids to be dried in the air should be nonhygroscopic and do not absorb moisture from the air[68]. The time taken to dry the sample depends on the properties of the residual solvents.

(2) **Desiccator dryness**

Desiccators with desiccants, such as silica gel, phosphorus pentoxide, calcium chloride, or calcium sulfate are commonly used to dry organic solids. The solids to be dried are normally contained in a beaker or a round-bottom flask, and then placed in the desiccator. Desiccator dryness can be carried out at atmospheric pressure or under a vacuum. Vacuum dryness can be accomplished more effectively by directly connecting the desiccator with a vacuum line.

(3) **Oven dryness**

Increasing the environment temperature can accelerate the drying rate. A solid can be dried more quickly by putting it in an oven if it is hygroscopic or has been recrystallized from water or a solvent with a high boiling point. It should be noted that the temperature of the

---

[66] 将干冰或液氮加入有机溶剂中作为冷却介质,可使体系降到更低的温度。将干冰加入丙酮中可使温度降至−78 ℃。将液氮加入丙酮中,最低温度可降至−196 ℃。

[67] 固体产物中少量水和有机溶剂的存在,会影响其重量、熔点、元素分析和光谱分析的准确性。

[68] 非吸湿性固体不会从空气中吸收水分,可以用空气干燥法干燥。

oven should be kept 20-30 ℃ below the melting or decomposition point of the solid[69]. Besides oven dryness, infrared lamp and microwave irradiator can also be used to dry thermostable solids at higher temperature[70].

(4) Other drying methods

Flowing gas can take away the solvent from the solids and can be used to dry organic solids. This method is common for drying air-sensitive solids that must be dried in an inert atmosphere, such as nitrogen or helium.

Water is sometimes removed from a solid by dissolving the solid in an organic solvent capable of forming an azeotrope with water. The water can be distilled out with the organic solvent by azeotropic distillation, and then recover the solids dissolving in the remaining solvents by removing the solvent[71].

2. Drying Organic Solutions

● **Principle**

There will always remain some water in the organic liquid in the case of extracting an organic liquid from an aqueous solution. The small amount of residue water in the solution must be removed prior to some further operation. For example, liquid organic compounds must be completely dried before distillation to avoid front fraction.

Drying is a necessary technique to remove a small amount of residue water from the target organic solution for further operations. The common methods for drying organic solutions can be classified into physical one and chemical one. The former includes absorption, fractional distillation and dehydration by molecular sieve, etc., while the latter is adopted to remove water by means of chemical reactions between drying agents. According to the mechanism of the reactions, chemical drying methods can be divided into the following two types:

① Anhydrous drying agents that can react with water irreversibly to form new substances, such as:

$$2Na + 2H_2O \longrightarrow 2NaOH + H_2 \uparrow$$

② Anhydrous drying agents that can react with water reversibly to form hydrates, such as:

$$CaCl_2 + nH_2O \rightleftharpoons CaCl_2 \cdot nH_2O$$

The latter one is probably the most used chemical drying method in organic laboratory. The routinely used drying agents are anhydrous inorganic salts, such as anhydrous calcium chloride, sodium sulfate or magnesium sulfate, potassium carbonate, etc[72]. The following fac-

---

[69] 烘箱干燥时，烘箱的设定温度应低于固体的熔点或分解点 20～30 ℃。

[70] 除了烘箱干燥外，红外灯和微波辐照器也可用于在较高温度下干燥耐热固体。

[71] 将固体溶解在能与水形成共沸物的有机溶剂中，通过共沸蒸馏将水与有机溶剂一起蒸出，除去剩余的溶剂。

[72] 实验室常用无水的无机盐作为干燥剂，如无水氯化钙、无水硫酸钠或硫酸镁、无水碳酸钾等。

tors should be considered for selecting a suitable dying reagent to dry an organic solution[73].

a. The drying reagent should be chemical inertness and cannot react with both the organic solution and any organic compounds dissolved in the solution.

b. The drying reagent cannot dissolve in the organic solution to be dried.

c. The drying reagent should have a large capacity for removing water.

d. The drying reagent can form hydrate with water quickly.

The amount of the required drying agent can be roughly calculated based on the amount of liquid to be dried and the capacity of the drying agent. Usually, the general amount of drying agent is 0.5-1 g per 10 mL liquid. In practice, the appearance of the following two phenomena in the Erlenmeyer flask containing the solution to be dried indicates the sufficient amount of drying agent[74]:

a. The newly added drying agent does not clump together and can flow freely in the solution when the flask is gently swirled.

b. The solution being treated with the drying agent turns clear and transparent from turbid.

- **Apparatus for Drying Organic Solution**

The operation of drying an organic solution with the drying agents is usually carried out in a dried Erlenmeyer flask. During the drying process, the flask should be closed using a stopper to prevent evaporation loss and the entry of moisture in the air.

- **Procedure for Drying Organic Solution**

① Transfer the solution to be dried in a dried Erlenmeyer flask.

② Add a small amount of drying agent directly into the organic solution and plug the stopper.

③ Swirl the Erlenmeyer flask gently and let it stand for 10-20 min.

④ Add more drying reagents and swirl the Erlenmeyer flask occasionally if the drying agent clumps together or if the solution still appears cloudy.

⑤ Repeat the above steps until the liquid appears clear and some of the drying agents are free-flowing.

- **Tips for Drying Organic Solutions**

① The exact amounts of drying agents depend on the amount of water in the organic solution to be dried, and vary from experiment to experiment. Careful observation turns to be the only way to determine the amount of drying agent to be used.

② Don't use too many drying agents while drying an organic solution, because drying agents may absorb the desired organic product along with the water[75].

---

[73] 选择干燥剂要考虑以下因素：①与被干燥的有机溶液或溶解在溶液中的物质不发生化学反应；②不能溶解在有机溶液中；③吸水量大；④与水形成配合物的速度快。

[74] 干燥剂的用量一般是10 mL液体加0.5~1 g的干燥剂。在实际操作中，当用于干燥用的锥形瓶中观察到以下现象时，表明干燥剂的量已经足够：①新加入的干燥剂不结块，可在锥形瓶底自由混动；②干燥的液体由浑浊变为澄清透明。

[75] 干燥剂的用量要适中，过多的干燥剂在吸水的同时，也会吸附产物。

③ Swirling Erlenmeyer flask is necessary during the dying process, which can increase the surface area for contact between the drying agents and liquid phases, thereby enhancing the drying speed❼⑥.

④ If an anhydrous inorganic salt is used for drying an organic solution, it must be completely removed by filtration before the dried liquid is to be distilled. Many hydrates will decompose with the loss of water at temperatures above 30-40 ℃ ❼⑦.

- **Questions**

① What are the characteristics of a good drying agent?

② Which will be a more effective drying agent, $MgSO_4$ or $MgSO_4 \cdot 7H_2O$? Explain your answer.

③ An anhydrous inorganic salt used for drying agents can be recycled after an experiment by heating its hydrates to drive off the water. Can the recycled drying agent be used continuously for another experiment? Explain the advantages and disadvantages of your proposal.

④ What are the disadvantages of adding too many drying agents to a solution? What can be done to avoid this situation?

2.3 Filtration

Filtration is an important technique used to separate solids from liquids. During the filtrating process, a porous barrier is used to allow the liquids to pass through and be collected in an appropriate receiver, while the solids should be retained on the barrier. Filter papers available in different sizes and porosities are commonly used to produce the barrier for filtration. The more porous the filter paper is, the faster the filtration rate becomes. Besides filter papers, cotton and glass wool can also be used as a barrier for filtration. They are more porous than filter paper, and are thus only used to separate the relatively large particles from liquids. The two most important kinds of filtration in organic laboratory are gravity filtration and vacuum filtration.

1. Gravity Filtration

In general, gravity filtration is usually adopted ① to remove solid drying agents from a final product solution to be distilled, and ② to remove solid insoluble impurities from a solution when the desired substance is in the solution. As the result of gravity filtration, the solids are discarded and the filtrate is collected for further use.

Gravity filtration is performed with a stemmed funnel and an Erlenmeyer flask. A round filter paper is folded to fit into the funnel and used as the barrier. The round filter paper in

---

⑯ 干燥时要不时地晃动锥形瓶，增大干燥剂和待干燥液体的接触面积，以加快干燥速度。

⑰ 许多无机盐的水合物在温度超过 30～40 ℃时，就会失水而分解。因而，当用无机盐作为干燥剂干燥液体化合物时，切记在蒸馏干燥的液体之前，必须通过过滤将干燥剂完全除去，以免干燥剂倒入蒸馏烧瓶中。

gravity filtration is often folded into a filter cone or a fluted filter. Fluted filter paper is more popular nowadays because it can provide a larger surface for liquid-solid separations and facilitate the gravity filtration[78]. Fluted filter can be easily prepared by creasing the round filter paper in half four times, then folding each of the eight newly presented sections inward, and finally opening the paper to form a fluted cone. The general procedures for gravity filtration are as follows:

① Fix the stemmed funnel onto an iron support using a ring support.

② Place an Erlenmeyer flask below the funnel as the receiving container. Then, adjust the height of the iron ring so that the tip of the funnel leans against the inner wall of the flask[79].

③ Select the proper filter paper to be used. The size of filter paper should be just slightly below the rim of the funnel.

④ Fold the filter paper into a fluted cone and insert it into the funnel.

⑤ Wet the filter paper with a small amount of the pure solvent and press it gently to make the paper adhere to the funnel[80].

⑥ Pour the mixture to be filtered onto the filter paper with the aid of a glass rod, and keep the liquid level in the funnel always 2-3 nm lower than the upper edge of the filter paper.

⑦ Add a few milliliters of the pure solvent to wash through the solids adhered to the filter paper after all the liquids have passed through the filter paper.

2. Vacuum Filtration

Vacuum filtration is used to separate solids from solutions rapidly, and collect the solid product from a reaction mixture or recrystallization mixtures. The equipment required for vacuum filtration includes Büchner funnel, filter flask, rubber adapter, filter paper, safety trap and water aspirator[81]. Typical apparatus for vacuum filtration is shown in Figure 2.1.

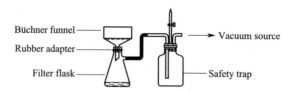

Figure 2.1 Apparatus for vacuum distillation

The vacuum source for vacuum filtration is usually a water aspirator. Herein, the Büchner funnel is attached to a filter flask by a rubber adapter that insures a tight seal between the filter flask and the Büchner funnel. The sidearm of the filter flask is connected to a

---

⑱ 槽纹滤纸（菊花滤纸）由于能提供较大的液固分离表面，更便于重力过滤，因而在常压过滤时被广泛采用。

⑲ 在漏斗下方放置一个锥形瓶作为接收器。调整铁圈的高低，使漏斗的尖端靠在锥形瓶的内壁。

⑳ 用少量纯溶剂润湿滤纸，轻压滤纸四周，使之紧贴在漏斗上。

㉑ 减压过滤常用的装置包括：布氏漏斗、抽滤瓶、橡皮塞、滤纸、安全瓶和水泵。

water aspirator using heavy-walled rubber tubing, and a safety trap is placed between the aspirator and the filter flask to prevent water from backing up into the filter flask in the case of a loss of water pressure suddenly[82]. Both the filter flask and the safety trap should be clamped to avoid breakage. The appropriate size of filter paper is crucial for a successful vacuum filtration. The filter paper lying flat on the plate of a Büchner funnel should just cover all the holes in the plate without curling up the side of the funnel[83]. Too large filter paper will cause some of the mixtures to flow over the unsealed edge and result in loss of products as well as contamination of the filtrate. The general procedures for a vacuum filtration are as follows:

① Place a piece of pre-cut filter paper with an appropriate size on the Büchner funnel.

② Wet the filter paper with a small amount of pure solvent.

③ Apply a vacuum to the system to pull the filter paper tightly over the holes in the Büchner funnel.

④ Swirl the mixture to be filtered and then transfer it into the funnel with the aid of a glass rod. Keep the liquid level in the funnel always lower than the upper edge of the funnel.

⑤ Transfer the remaining solids in the container by washing them with some of mother liquor in the filter flask.

⑥ Open the stopcock on the safety trap bottle first, then add a small amount of solvent to drench the solids if the solids on the Büchner funnel need to be washed. In the case of liquid dripping from the tip of the Büchner funnel, close the stopcock on the safety trap and pump the solids dry[84].

⑦ Open the stopcock on the safety trap to release the vacuum before the water aspirator is turned off when all the liquids has been drawn through the Büchner funnel, thereby avoiding the water in the water aspirator flowing back into the filter flask[85].

Vacuum filtration can also be used for hot filtration to remove the insoluble impurities and decolorize activated carbon in recrystallization. Rapid hot filtration effectively prevents premature crystallization of the crystals in the filter paper and the filter flask. For a hot filtration, it is desirable to preheat both the Büchner funnel and the filter flask by hot eater. Meanwhile, the filter paper also needs to be wet with hot pure solvent in advance[86].

2.4 Simple Distillation and Boiling Point

Principle

When liquid is heated, its vapor pressure increases as the temperature increases. The

---

[82] 在水泵和抽滤瓶之间设置了一个安全瓶，以防止压力突然变小时，水泵中的水倒吸到抽滤瓶内。

[83] 滤纸的大小对减压过滤至关重要。平放在布氏漏斗上的滤纸应该能刚好盖住漏斗上所有的孔，而不能卷起抵住漏斗的一侧。

[84] 若需对固体进行洗涤，应先打开安全瓶上的开关，再用少量溶剂淋洗布氏漏斗上的固体。当有液体自布氏漏斗下端滴下时，再关闭安全瓶上的开关，将固体抽干即可。

[85] 抽滤结束后，应先打开安全瓶上的开关，然后再关闭水泵，以免造成水流倒吸。

[86] 在进行减压热过滤时，事先要将布氏漏斗和抽滤瓶放在热水中预热，滤纸也要用热的溶剂打湿。

boiling point is defined as the temperature when the vapor pressure above the heated liquid equals to the external pressure[87]. Every kind of pure compound has a fixed boiling point at given pressure. The boiling point is dependent upon the external pressure, and the boiling point is usually lower at a place where the atmospheric pressure is lower. The boiling point of a pure compound reported in the literature is referred to one atmospheric pressure.

Simple distillation is the operation consisting of boiling a liquid to gaseous vapors and then condensing the vapors back to the liquid state. The liquid collected in a receiving vessel during the distillation is usually called fraction(or distillate)[88]. When the distillation is of a pure liquid, it will boil at its boiling point and the distillate is received at the fixed point with narrow temperature range. Conversely, a liquid mixture usually boils over a fairly wide range. The boiling point and the composition of the distillate will change during the distillation process. The component with low boiling point is distilled out first and that with high boiling point is distilled out later, leaving the non-volatile components[89] in the distillation flask.

In summary, distillation has the following implications in organic laboratory[90]:

① Determining the boiling point of a pure liquid compound.

② Determining the purity of a liquid compound.

The boiling range of a pure liquid substance is about 0.5-1 ℃, while a liquid mixture will have a large boiling range. The purity of a liquid can be determined by the temperature range of its boiling point. However, this approach is limited because a binary or ternary azeotrope[91], which is formed by an organic compound with other components, also has a definite boiling point with narrow range.

③ Separation and purification of a liquid mixture.

Simple distillation is frequently used in organic synthesis to isolate the final product from a liquid mixture. For example, most of the organic solvents used for the reaction or its work-up are highly volatile with relatively lower boiling points, so they can be removed easily from the organic compounds with higher boiling points by simple distillation. It is worth noticing that simple distillation can work well only when the difference between the boiling point of each pure component in the liquid mixture is greater than 30 ℃.

- **Apparatus for Simple Distillation**

Typical simple distillation apparatus is shown in Figure 2.2.

The important pieces include[92]: distillation flask, distillation head, thermometer adapter, thermometer, condenser, bent adapter and receiving flask(Erlenmeyer flask).

---

[87] 液态有机物的蒸气压随温度升高增大，当其蒸气压达到大气压时，液体开始沸腾，此时的温度称为液体的沸点。

[88] 馏分。

[89] 不挥发的组分。

[90] 蒸馏的意义有以下三个方面：①测定某纯液态物质的沸点；②判定液体化合物的纯度；③分离和提纯液体混合物。

[91] 二元或三元共沸物。

[92] 主要仪器包括：蒸馏烧瓶、蒸馏头、温度计套管、温度计、接引管、接收瓶（锥形瓶）。

- **Procedures for Simple Distillation**

① Clamp the distillation flask to the iron stand, then assemble the simple distillation apparatus (Figure 2.2) following the "bottom-to-top" and "left-to-right" sequence.

② Insert a thermometer into the distillation head through the thermometer adapter. Make sure the position of the top of the mercury thermometer bulb is levelled with the bottom of the sidearm of the distillation head❸.

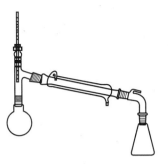

Figure 2.2 Apparatus for simple distillation

③ Attach rubber tube to the condenser (one for water in and the other for water out) and clamp the condenser to a second ring stand.

④ Connect the condenser with the sidearm of distillation head and attach a bent adapter to the condenser. Make sure that the cooling water enters from the bottom of a condenser and exits at the top❹.

⑤ Place a receiving flask beneath the adapter.

⑥ Add the liquid to be distilled into the distillation flask, then add a couple of zeolites prior to the distillation to produce smooth bubbling and prevent boiling over. Do not add zeolites to a boiling liquid❺.

⑦ Introduce tap water into the condenser. Heat the flask slowly with a heating mantle, and then gradually increase the heating power, making the liquid boiling.

⑧ Adjust the heating power, maintaining distillation rate at 1-2 d/s. Record the temperature when the first distillate dropping into the receiving flask. Control the power properly and collect all distillate over the range of desired temperature.

⑨ Remove the heating source, stop the cooling water, and disassemble the apparatus following the "top-to-bottom" and "right-to-left" sequence.

- **Tips for Simple Distillation**

① The volume of liquid in the distillation flask should not exceed two-thirds of the distillation flask❻. Too much liquid in the flask will cause bumping and contaminate the distillate.

② Air condenser should be used when a liquid with high boiling point (>140 ℃) is to be distilled❼.

③ Attach the tube before installing the condenser to the system. Make sure that the rubber tube is securely fastened to condenser.

④ Make sure that all connections in the apparatus are connected closely before turning

---

❸ 温度计水银球的上端与蒸馏头支管口的下端在同一平面。
❹ 从冷凝管的下端进水，上端出水，并且出水口要在冷凝管上方，以确保冷凝管的夹层中充满水。
❺ 加热前在蒸馏烧瓶中加入沸石，不要向沸腾的溶液中加入沸石，以免造成溶液暴沸。
❻ 蒸馏液体的体积占蒸馏烧瓶容量的 1/3~2/3。
❼ 蒸馏时，若被蒸馏物质的沸点高于 140 ℃，应使用空气冷凝管。

on the heating mantle[98].

⑤ Be sure that the whole system is open to the outside atmosphere, otherwise an explosion may occur as the pressure builds up within the system[99].

⑥ Control the heating rate to ensure that distillation continues steadily at a rate of 1-2 d/s.

⑦ Don't distill to dryness to prevent any experiment accidents. Turn off heating when there is only 2-3 mL of liquid remained in the distillation flask.

⑧ When the distillation is carried out with a low-boiling liquid, place the receiving flask into an ice-water bath to prevent evaporation and to ensure that the distillate is completed condensed[100].

- **Questions**

① What possible hazards might occur if a distillation flask is heated to dryness?

② What is the role of zeolites when it is added to a liquid to be heated to boiling?

③ What is the accurate position of the thermometer for simple distillation? Explain the effect on the observed temperature reading if the thermometer is placed higher or lower.

④ At 100 ℃, the gaseous vapor pressures of toluene and water are 564 torr and 760 torr (1 torr = 133 Pa), respectively. Please speculate which one has the higher normal boiling point.

2.5 Fractional Distillation

- **Principle**

A mixture of two or more volatile liquid components cannot be separated into pure component by simple distillation if the boiling points of these substances are close to each other. As mentioned earlier, simple distillation is only applicable when the difference of boiling points between the different components is more than 30 ℃.

The technique of fractional distillation is usually used for separation and purification of liquid organic compounds with a small boiling point difference. The principle and apparatus of simple distillation and fractional distillation are quite similar, except that a fractionating column is added between the distillation flask and the distillation head in fractional distillation apparatus[101](Figure 2.3). There are many kinds of fractionating column. Among them, the Vigreux column is most commonly used in organic laboratory.

As the vapor of a liquid mixture from the distillation flask rises up through the fractionating column, some of it will condense upon heat exchange with the cooler surface of the col-

---

[98] 加热前要确保所有的磨口都连接紧密。

[99] 常压蒸馏的整套体系必须在接引管处与大气相通，以免压力过大造成爆炸。

[100] 蒸馏低沸点化合物时，应将接受瓶放在冰水浴中冷却，以确保馏出物的充分冷凝，防止挥发。

[101] 分馏的基本原理和装置与蒸馏相类似，所不同的是在蒸馏烧瓶和蒸馏头之间加了一根分馏柱，利用分馏柱实现了"多次重复"的蒸馏过程。

umn. There is heat exchange between the rising vapor and descending condensate[10], which makes the condensate revaporized partially. As a result of the cycle of condensation-revaporization, there are more components with lower boiling-point in the rising vapor and more components with higher boiling-point in the condensate[11]. The cycle of condensation-revaporization can be repeated for many times in a long fractionating column, which leads to the vapor phase becoming increasingly richer in the lower boiling-point component and going higher and higher in the column, while the condensate correspondingly becomes increasingly richer in the higher boiling-point component and goes back to the flask. If the fractional distillation is efficient enough, the vapor at the top of the column would be close to pure component with lower boiling point. Meanwhile, the higher boiling-point component almost remains in the flask. In fact, each condensation-revaporization cycle is actually a simple distillation, so the repetitive cycles in the column is equivalent to performing simple distillation for multiple times[12]. Thus, a better separation of components with small difference in boiling point is achieved through fractional distillation.

In summary, fractional distillation uses a fractionating column to achieve successive simple distillation. Obviously, fractional distillation is more effective than simple distillation and is applicable to separating a mixture of two or more liquid components with small difference ($< 30\ ℃$) in boiling points.

- **Apparatus for Fractional Distillation**

Typical fractional distillation apparatus is shown in Figure 2.3.

The important pieces include: distillation flask, fractionating column, distillation head, thermometer adapter, thermometer, condenser, bent adapter and receiving flask.

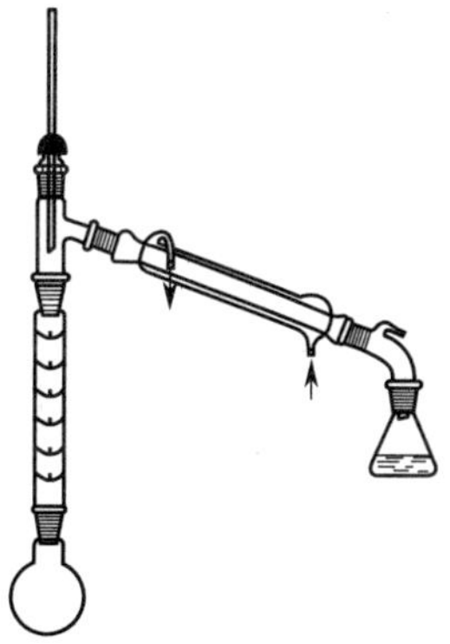

- **Procedures for Fractional Distillation**

① Select a fractionating column, assemble the fractional distillation apparatus (Figure 2.3) following the "bottom-to-top" and "left-to-right" sequence.

② Add the liquid to be distilled and a couple of zeolites into the distillation flask. Make sure that all joints are connected well.

③ Introduce tap water into the condenser and heat the flask slowly with a heating mantle.

Figure 2.3 Apparatus for fractional distillation

④ Adjust the heating power, maintaining distillation at a rate of 1d/2s-1d/3s and record the temperature of different kinds of distillates. Remove the heating source, stop the cooling water.

⑤ Disassemble the apparatus in "top-to-bottom" and "right-to-left" sequence.

---

⑩ 冷凝液。

⑪ 冷凝-再气化的循环，使得上升的蒸气中有更多沸点较低的组分，而冷凝液中沸点较高的组分较多。

⑫ 每一次气化-凝结的循环实际上就是一次简单的蒸馏。蒸气就是这样在分馏柱内反复地进行着气化-冷凝的过程，或者说在重复循环地进行着若干次的简单蒸馏。

• **Tips for Fractional Distillation**

The tips given for simple distillation also apply to fractional distillation. Additional points are given below:

① The apparatus for fractional distillation is much taller than that for simple distillation. A high clearance is necessary on the lab bench.

② More heating power should be applied for fractional distillation than those applied for simple distillation.

③ It is often useful to wrap the fractionating column with asbestos to ensure the vapors of distillate reach the top of condenser continuously[⑩⑤].

④ Control the speed of heating. If the rate of distillation is too high, the purity of distillate will decrease. If the distillation rate is too low, the reading of thermometer may fluctuate up and down[⑩⑥].

⑤ For multicomponent fractional distillation, it is important to observe the change of thermometer reading. It might be a good time to change receiving flask for another distillate when the temperature observed in the still head decreases during distillation[⑩⑦].

• **Questions**

① What are the similarities and differences between the fractional distillation and simple distillation?

② What will affect the efficiency of fractional distillation?

③ When a fractional distillation is carried out with a binary mixture, what might cause the thermometer reading suddenly begin to decrease under the constant heating condition? Why is that?

2.6 Steam Distillation

• **Principle**

The process that a mixture of one or more organic liquids with water co-distill is called steam distillation. It is another kind of distillation that is especially useful to separate the high boiling-point volatile organic components, which are immiscible or nearly immiscible with water, from inorganic salts or other nonvolatile organic compounds, such as resin-like or oily substance.

When a mixture of organic compound A and water is heated together, according to the law of partial pressure, each component of the mixture contributes its vapor pressure to the total pressure as expressed in the following equation:

$$p_{total} = p_{water} + p_A$$

---

⑩⑤ 为确保蒸气能到达分馏柱顶端，可用石棉绳将柱子包裹起来。

⑩⑥ 分馏时，应控制加热的速度。分馏速度太快，产品纯度会下降；分馏速度过慢，馏出温度易上下波动。

⑩⑦ 多组分液体混合物的分馏，观察温度计读数的变化十分关键。当计温度度数下降时，应准备更换接收瓶，接收另一种馏分。

When the total vapor pressure above the mixture is equal to external pressure, the mixture of compound A and water starts boiling and co-distilling. Obviously, the boiling point of the mixture will be lower than either of water or compound A. This means that the organic compound A can be distilled with water at a boiling point which is much lower than that of pure compound A[9].

In the steam distillation, the weight ratio of organic compound (w_A) and water (w_{H_2O}) can be calculated from the following equation:

$$\frac{w_A}{w_{H_2O}} = \frac{p_A M_A}{p_{H_2O} M_{H_2O}}$$

Steam distillation is one of the commonly used methods for separation and purification of liquid mixtures, but it is limited and can only be applied to the following situations[10]:

① High boiling-point organic compounds which are heat-sensitive and would therefore decompose at higher temperature.

② A mixture containing inorganic or other organic solid salts, which is not suitable for distillation, filtration and extraction.

③ A mixture containing large amounts of resin-like substances or non-volatile impurities which is hard to isolate by simple distillation, filtration and extraction.

Although the conditions used in steam distillation is relatively mild, it is least frequently used comparing with other types of distillation. The substances to be separated by steam distillation should have the following characteristics[11]:

① The compounds are insoluble or nearly insoluble in water.

② The compounds cannot react with water.

③ The compounds do not decompose even after prolonged contact with steam or hot water.

④ Have a certain vapor pressure (>10 mmHg) at 100 ℃.

- **Apparatus for Steam Distillation**

Typical steam distillation apparatus is shown in Figure 2.4.

The important pieces include: round-bottom flask, T-shaped glass tube, safety column, three-neck distillation flask, distillation head, thermometer adapter, thermometer, condenser, bent adapter and receiving flask.

- **Procedures for Steam Distillation**

① Set up the steam distillation apparatus (Figure 2.4) following the "bottom-to-top" and "left-to-right" sequence.

---

⑨ 显然，混合物的沸点要低于水或有机物 A 的沸点，即有机物 A 通过水蒸气蒸馏可以在比其沸点低的温度下和水一起蒸馏出来。

⑩ 水蒸气蒸馏适用于以下几种情况：①在高温条件易分解的高沸点有机化合物；②含有较多固体混合物，用简单蒸馏、过滤或萃取等方法难以分离；③混合物中含有大量树脂状或不挥发性杂质，采用简单蒸馏、过滤或萃取等方法难以分离。

⑪ 可以通过水蒸气蒸馏分离的化合物应具有以下特征：①不溶或难溶于水；②和水不发生反应；③即使长时间接触蒸气或热水也不会分解；④在 100 ℃左右该物质的蒸气压至少在 10 mmHg 以上。

② Add the mixture to be distilled to the three-neck distillation flask. Make sure that all joints are connected well.

③ Open the screw clamp on the T-shaped glass tube connecting with the vapor-generator and distillation flask.

④ Introduce tap water into the condenser and heat the vapor-generator with a heating mantle.

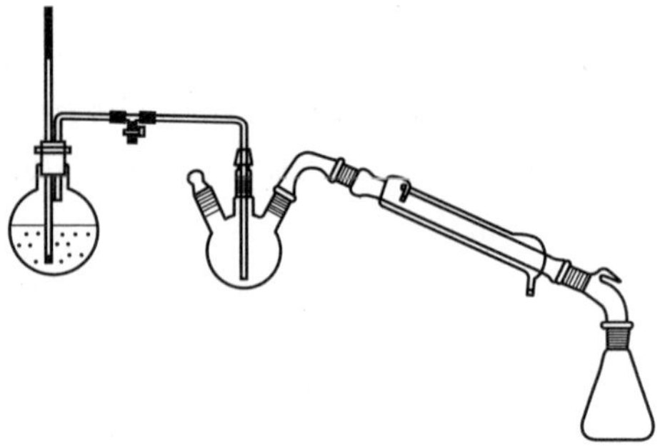

⑤ Close the screw clamp on the T-shaped glass tube to let steam enter the distillation flask when a large amount of steam is generated from the vapor-generator[⑪].

Figure 2.4 Apparatus for steam distillation

⑥ Adjust the heating power, maintaining distillation at a rate of 2-3 d/s.

⑦ Continue the distillation until oily substances no longer appear in the distillate[⑫].

⑧ Open the screw clamp on the T-shaped glass tube so that the whole system is connected to the atmosphere before it stop heating[⑬].

⑨ Remove the heating source, stop the cooling water, and disassemble the apparatus in "top-to-bottom" and "right-to-left" sequence.

- **Tips for Steam Distillation**

Some tips given for simple distillation also apply to fractional distillation. Additional points are given below:

① Water in the vapor-generator should be no more than 2/3 full to avoid boiling water rushing out of the generator.

② The glass tube for introducing gas should be immersed in the solution and be as close as possible to the bottom of three-neck distillation-flask.

③ The mixture to be distilled should be no more than 1/3 of volume of the three-neck distilling-flask. Heat the three-neck distillation flask with a small flame during the steam distillation to avoid the accumulation of condensed water in it[⑭].

④ Watch the water level in the safety column frequently during steam distillation. If the water level in the column suddenly rises, open the screw of the T-shaped tube immediately and remove the heating mantle. Inspect the system and fix the problem before carrying on[⑮].

- **Questions**

① Describe the principle of steam distillation.

② Does the boiling point of a mixture ever exceed 100 ℃ during steam distillation? Ex-

---

⑪ 当有大量蒸气产生时，关闭 T 形管上的螺旋夹，使蒸气均匀地进入蒸馏烧瓶。

⑫ 当馏出液不再含有油滴时，即可停止水蒸气蒸馏。

⑬ 停止蒸馏时，必须先打开 T 形管上的螺旋夹，待体系和大气相通后，再停止加热。

⑭ 被蒸馏液体的体积不要超过蒸馏烧瓶容积的 1/3。为了防止水蒸气在蒸馏烧瓶中过多冷凝，水蒸气蒸馏过程用可用小火加热蒸馏烧瓶。

⑮ 水蒸气蒸馏过程中，要注意观察安全管中水柱的情况。若有异常，立刻打开 T 形管上的螺旋夹，移去电热套。检查系统并排除故障后方可继续。

plain your answer.

③ Explain why 1-propanol with the boiling point of 97 ℃(760 mmHg) cannot be purified by steam distillation?

④ Explain why a water-insoluble organic compound whose vapor pressure is negligible at 100℃ can not be purified by steam distillation?

⑤ How to determine the end-point of steam distillation?

2.7 Vacuum Distillation

• **Principle**

Some organic compounds, which decompose at temperatures below their boiling points, cannot be purified by simple distillation at ordinary pressures. In practice, these substances can be distilled under reduced pressure. Such distillation carried out at the pressures below atmospheric pressure is called vacuum distillation⑯. In general, the compounds that boil above 200 ℃ at normal atmospheric pressure are usually purified by vacuum distillation to avoid chemical changes which may occur at high temperature.

When the vapor pressure above the liquid equals to the external pressure, the liquid begins to boil. If the pressure above the liquid is reduced, the liquid will boil at a lower temperature. As a rough estimate, a drop in pressure of 1 mmHg lowers the boiling point of an organic liquid by about 1 ℃ when the vacuum distillation is performed at 10-25 mmHg. In general, if the external pressure is reduced to 20 mmHg, the boiling point is 100-120℃ lower than the boiling point at atmospheric pressure⑰.

In practice, it is very difficult to calculate the exact boiling point of a compound at any given pressure other than 760 mmHg. A nomograph can provide a good way of estimating the boiling point of a compound at a certain pressure(Figure 2.5).

There are three scales in the nomograph. Scale A, B and C present respectively the boiling point at certain pressure, the boiling point at one atmospheric pressure, and the actual pressure over the distillation system. For example, if the boiling point of a compound is 200 ℃ at 760 mmHg and the vacuum distillation is performed at 20 mmHg, the approximate boiling point of this compound at 20 mmHg is found by connecting 200 mmHg in scale B with 20 mmHg in scale C using a ruler, and extending the line, then observing its intersection with scale A. The reading 90 ℃ at the intersection is the approximate boiling point of this compound at 20 mmHg.

The standard apparatus for vacuum distillation generally consists of four parts: distillation system, vacuum source, protection and pressure-measuring equipment(Figure 2.6)⑱. A vacuum can be obtained in the organic laboratory by using either a mechanical pump or a wa-

---

⑯ 在较低压力下进行的蒸馏操作，称为减压蒸馏。

⑰ 当减压蒸馏在 10~25 mmHg 之间进行时，压力每下降 1 mmHg，沸点将降低 1 ℃。一般情况下，当体系压力降低至 20 mmHg，沸点将比常压下的沸点降低 100~120 ℃。

⑱ 减压蒸馏装置一般由蒸馏、抽气、保护和测压四部分组成。

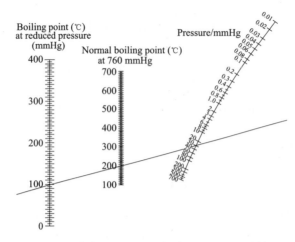

Figure 2.5　Nomograph for estimating boiling points at different pressures

ter aspirator[⑱]. Mechanical pumps can easily produce pressures of less than 0.5 mmHg. The pressure obtained with a water aspirator is approximately 15 mmHg at 25 ℃. A pressure-measuring device is usually placed between the vacuum source and the vacuum adaptor to show the actual pressure of the distillation system. When a mechanical pump is used as the vacuum source, a cold trap and several absorption columns must be placed between the distillation system and the pump to protect the pump. The trap is used to condense the volatile substances with low boiling points and thus to prevent them from going into the mechanical bump, which would reduce the efficiency of the pump[⑳]. The three absorption columns filled with anhydrous calcium chloride, sodium hydroxide and activated carbon in sequence can absorb water, acidic substances and organic hydrocarbons respectively to protect the pump[㉑]. A safety flask is needed between the vacuum adaptor and the trap to close the system from the atmosphere and to release the vacuum as the system cools down after the distillation[㉒]. When a water aspirator is used as the vacuum source, a solvent trap bottle or flask must be used to prevent the back flow of water from entering the receiver.

In summary, vacuum distillation can be applied to purify the compounds that boil at higher temperature(＞200 ℃) at atmospheric pressure or compounds which may easily decompose at the temperature below their atmospheric boiling points[㉓]. Vacuum distillation provides a method to allow these compounds to boil at a lower temperature and thus be distilled without decomposition.

- **Apparatus for Vacuum Distillation**

Typical vacuum distillation apparatus is shown in Figure 2.6.

---

⑱　水泵。
⑲　冷阱用来冷凝低沸点的挥发性物质，以免其进入油泵，造成油泵效能的降低。
⑳　三个吸收塔依次装有无水氯化钙、氢氧化钠和活性炭，分别用来吸收水分、酸性物质和有机烃类物质。
㉑　在接引管和冷阱之间还必须接上一个安全瓶，瓶上的两通活塞可以起到调节体系压力和放气的作用。
㉒　减压蒸馏特别适用于那些沸点高（＞200 ℃），且在常压下蒸馏未达到沸点就已受热分解、氧化或聚合物质的提纯和分离。

The important pieces include: distillation flask, capillary bubbler, Claisen head, thermometer adapter, thermometer, condenser, vacuum adaptor and receiving flask, safety flask, cold trap, absorption columns and vacuum pump.

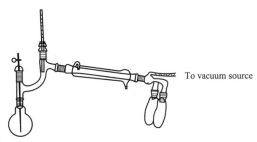

Figure 2.6 Apparatus for vacuum distillation

- **Procedures for Vacuum Distillation**

① Assemble the vacuum distillation apparatus (Figure 2.6) following the "bottom-to-top" and "left-to-right" sequence.

② Close the screw clamp at the top of the capillary bubbler and open the pressure release valve of safety flask. Then, apply vacuum to the distillation by closing the pressure release valve of safety flask to test the seals of the apparatus[124].

③ If the desired low pressure can not be obtained, check every part of the system to find the leaks and seal them.

④ After checking the empty system, break the vacuum by opening the pressure release valve of safety flask slowly.

⑤ Add the liquid to be distilled to a distillation flask, apply a vacuum to the system by closing the pressure release valve of safety flask slowly.

⑥ Adjust the screw clamp of the capillary bubbler until steady bubbles appear from the bottom of the capillary.

⑦ Introduce tap water into the condenser. Heat the distillation flask after the vacuum is stable and has reached the desired level.

⑧ Adjust the heating power, maintaining distillation at a rate of 1-2 d/s.

⑨ Remove the heat source when the distillation has completed.

⑩ After the apparatus is cooled to room temperature, slowly open the screw clamp at the top of the capillary bubble and the pressure release valve of safety flask to introduce air to the system. Then, turn off the vacuum pump[125].

- **Tips for Vacuum Distillation**

The tips given for simple distillation also apply to fractional distillation. Additional points are given below:

① The glassware used for vacuum distillation must be vacuum resistant. Erlenmeyer flasks can not be used as receiving flask.

② Apply vacuum grease to the joints of the glassware to obtain better vacuum and to protect the system from open air.

③ The capillary bubble is used instead of zeolites to ensure that the distillation is stable by providing a steady stream of very small bubbles. The bubble is inserted into the distil-

---

[124] 旋紧毛细管上的螺旋夹，打开安全瓶上的开关，开泵抽气。缓慢关闭安全瓶上的旋塞，测试系统的气密性。

[125] 仪器冷却至室温后，慢慢打开毛细管顶部的螺旋夹和安全瓶的旋塞，使系统与大气相通，然后关闭真空泵。

lation flask through a thermometer adapter and the bottom of the bubble should be just above the bottom surface of the flask. A short rubber tube with a screw clamp is installed on the top of capillary bubble to control the rate of bubbling and the amount of air entering the system[26].

④ A Claisen connecting adapter rather than a normal distillation head is always used to reduce the possibility of the liquid bumping up into the condenser in a vacuum distillation[27].

⑤ Three-way distillation receiving adapter is useful setup for vacuum distillations when more than one distillate will be collected along the distillation course. This adapter can be used to introduce different kinds of distillate into appointed receiving flasks by rotating it while not disconnecting the vacuum and changing the receiving flask(Figure 2.7).[28]

⑥ The volume of liquid in the distillation flask should not exceed 1/2 volume of the distillation flask.

⑦ Most of the organic solvents boil below room temperature under vacuum. The solvents in the mixtures to be distilled should be removed before vacuum distillation to prevent uncondensed organic vapor from entering the pump[29].

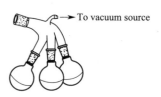

Figure 2.7　Three-way distillation receiving adapter

• **Questions**

① Describe the principles of vacuum distillation.

② What are the differences between the apparatus of vacuum distillation and simple distillation?

③ What should be noticed at the beginning and the end of vacuum distillation?

④ The boiling point of certain compound is 200 ℃ at atmospheric pressure(760 mmHg). Use the nomograph to determine its boiling point at 20 mmHg.

2.8　Extraction and Washing

• **Principle**

Liquid-liquid extraction is one of the most commonly used methods in organic laboratory to isolate the desired substance from a liquid mixture. The process of a liquid-liquid extraction is based on the different solubilities of a substance between two immiscible solvents, which assists in selectively transporting the substance from one solvent to another during an

---

㉖ 在减压蒸馏操作中一般用毛细管代替沸石，通过毛细管提供稳定的小气泡来确保蒸馏的平稳进行。毛细管通过温度计套管插入蒸馏烧瓶中，毛细管的下端离烧瓶底部约 1～2cm。其上端安装一根带螺丝夹的短橡胶管，以控制气泡的大小和进入系统的空气量。

㉗ 用克氏蒸馏头代替普通蒸馏头，避免减压蒸馏时瓶内的液体直接冲入冷凝管中。

㉘ 如果减压蒸馏过程中需要收集多个馏分而又不能断开真空更换接收瓶时，应用多尾接液管代替普通的接液管。此时只需转动多尾接液管，就可以使不同的馏分进入指定的接收瓶中。

㉙ 大多数的有机溶剂在减压的条件下，会在低于室温的温度沸腾。为了防止没有来得及冷凝的溶剂蒸气进入油泵，在减压蒸馏前应先通过普通蒸馏将有机溶剂除去。

extraction, thus achieving the goal of separation or purification❶. Liquid-liquid extraction is usually carried out by using a solvent pair of water and a less polar organic solvent. Most organic compounds prefer staying in a more soluble organic phase, whereas ionic or rather polar compounds prefer dissolving in water. In this case, the organic components can be extracted from an aqueous solution by an appropriate organic solvent after two layers of the water phase and the organic phase are separated in a separatory funnel.

At a given temperature and pressure, the concentration ratio of a compound X in two immiscible solvents, such as water and an organic solvent, is a constant, which is called the distribution coefficient(K). This relationship is known as the distribution principle, and can be described by the following equation❷:

$$K = \frac{\text{concentration of X in organic solvent}(g/mL)}{\text{concentration of X in water}(g/mL)}$$

The distribution principle is the basis of an extraction, and the distribution coefficient (K) can be used to evaluate the efficiency of an organic extracting solvent, with a larger value of K indicating a more efficient organic solvent. If the distribution coefficient K is large enough, a single extraction may be sufficient to extract the organic compound from water into the organic solvent. However, multiple extractions are required for most organic solvents to achieve the desired result.

In general, the fraction of a compound remaining in the original solvent after multiple extractions using an organic solvent follows the following equation:

$$W_n = W_0 \left(\frac{KV}{KV+S}\right)^n$$

Where,

W_0——the total weight of the compound to be extracted;

W_n——the weight of the compound remains in water after multiple(n) extractions;

K——the distribution coefficient;

V——the volume of the original solution;

S——the volume of the organic solvent in each extraction.

The above equation indicates that in the case of a given amount of organic solvent, a series of extractions using small volumes of solvents once are more efficient than a single large-volume extraction. However, considering the cost and time, it does not mean more times of extraction is always better. Generally, three times of extraction should be appropriate❸.

The extraction efficiency is also related to the properties of extraction solvents. A suitable organic solvent is the key to an effective liquid-liquid extraction. Methylene chloride, diethyl ether, ethyl acetate and hexane are the mostly used organic solvents for liquid-liquid ex-

---

❶ 萃取是利用某一物质在两种不相溶的溶剂中溶解度或分配比的不同，使该物质从一种溶剂转移到另一种溶剂中，从而达到提纯或分离的目的。

❷ 分配定律是指在一定温度、一定压力下，一种物质在两种互不相溶的溶剂中的分配浓度是个常数，即分配系数。

❸ 由此可见，用相同用量的溶剂分 n 次萃取比一次萃取的效果好。但并非萃取次数越多越好。考虑到时间、成本等因素，一般以三次萃取为宜。

traction. To select a suitable organic extraction solvent, the following general guidelines should be obeyed⑱:

① The extraction solvent is immiscible with the original solvent(usually water).

② The extraction solvent should have a favorable distribution coefficient.

③ The extraction solvent should be volatile and can be easily removed from the compound after extraction.

④ The extraction solvent should be inert, less toxic, and economical.

The term "washing" follows the same principle as "extraction". Extraction is the process of separating a desired compound from a mixture, while washing usually means removing impurities from the desired substance⑲. For example, when an organic phase contains inorganic acids or bases, washing the organic phase with aqueous solution of a base or an acid can remove the acidic or basic impurities from the organic solution.

Overall, liquid-liquid extraction and washing are common purification techniques in organic laboratory, which are usually applied in the following cases:

① To collect the product from the aqueous phase of a organic reaction;

② To isolate the organic substances from the aqueous solution;

③ To remove the inorganic salts, basic and acidic impurities from the organic phase.

- **Apparatus for Liquid-liquid Extraction and Washing**

Liquid-liquid extraction and washing are all accomplished in a pear-shaped separatory funnel, which is fragile and must be held securely when being shaken and vented during the operation of extraction or washing⑳. A separatory funnel is always held in the following way(Figure 2.8):

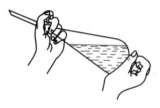

Figure 2.8　The application of separatory funnel

- **Procedures for Extraction and Washing**

① Detect the leakage of the separation funnel before use.

② Attach an iron ring to an iron stand. Place a separatory funnel on the ring and an Erlenmeyer flask under the separatory funnel.

③ Pour the solution to be extracted into the separatory funnel with the stopcock closed. The total volume of the solution to be extracted should be less than 3/4 that of the separatory funnel used.

④ Add the extraction solvent to the separatory funnel from the top. The volume of the extraction solvent should be nearly equal to that of solution to be extracted, and should be divided into 2-3 portions.

⑤ Place the stopper securely, remove the funnel from the iron ring and hold the funnel

---

⑱ 选择合适萃取溶剂的一般要求：①与原溶剂不相溶；②具有合适的分配系数；③易于与被提纯物质分离；④化学性质稳定，毒性小，经济适用。

⑲ 洗涤和萃取的基本原理一样，从原来的溶剂中提取出所需要的组分的过程称为萃取，从混合物中除去不需要的少量杂质的操作称为洗涤。

⑳ 分离漏斗易碎，振摇或需要放气时，必须拿好握紧。

in the way shown in Figure 2.8.

⑥ Invert the funnel slowly and open the stopcock to release any pressure build-up⑯.

⑦ Close the stopcock, shake the mixture for a while, and then vent the funnel slowly to release the pressure by opening the stopcock.

⑧ Repeat the shaking, inverting and venting processes 3-4 times.

⑨ Stand the funnel on the iron ring until the two layers have completely separated with a sharp demarcation line between the two phases.

⑩ Remove the stopper, open the stopcock to drain the bottom layer into an Erlenmeyer flask and pour the remaining top layer out of the funnel through the top into another Erlenmeyer flask⑰.

- **Tips for Liquid-liquid Extraction and Washing**

① Lightly grease the surface of ground-glass stopcock of a separatory funnel to avoid sticking or leaking.

② Vent the separatory funnel to avoid a buildup of pressure, especially when a low-boiling extraction solvents is used, or when a dilute sodium carbonate or bicarbonate solution is used to wash the organic phase containing traces of an acid. The separatory funnel cannot be aimed to persons nearby during venting to avoid incidents⑱.

③ The following methods can help you to determine which layer is the organic layer and which is the aqueous layer⑲:

a. Compare the density of extracting solvent and water. Usually, the one with a larger density will be on the bottom.

b. Add 1-2 mL of water to the separatory funnel, the layer on which the water drops stay must be the aqueous layer.

c. Take out 1-2 drops of liquid from one of the layers and put onto a watch glass, and then add several drops of water. The layer must be organic layer if the liquid on the watch glass appears cloudy.

④ An emulsion may be formed if the two layers are not separated clearly and the entire mixture has a milky appearance, or there may be a third milky layer between the aqueous and the organic phases. There are two things to do to break it⑳:

a. Stir the mixture with a glass rod, and let it stand for several minutes.

b. Add a small amount of saturated sodium chloride solution or solid sodium chloride to

---

⑯ 将分液漏斗倒置，打开活塞放气。

⑰ 拿掉上面的玻璃塞，将活塞旋开，下层的液体自活塞处放出，上层的液体从分液漏斗的上口倒出。

⑱ 当使用低沸点的萃取溶剂时，或使用稀的碳酸钠或碳酸氢钠溶液来洗去有机相中含有的微量酸时，要注意及时放气，避免压力的积聚。在放气过程中，不要将分液漏斗对准附近的人，以避免发生意外。

⑲ 可以用以下的方法来确定哪一层是有机相，哪一层是水相：a. 比较萃取溶剂和水的密度。通常情况下，密度大的一方会在底部；b. 在分液漏斗中加入1～2 mL水，水滴停留的层那一层就是水相；c. 从其中一层中取出1～2滴液体，滴在表面皿上，然后加入几滴水。如果表面皿上的液体出现浑浊，则该层是有机相。

⑳ 如果分层现象不明显，整个混合物呈现乳白色，或者在水相和有机相之间出现了一层乳白色液体，这种现象称为乳化。有两种方法可以帮助我们消除乳化现象：a. 用玻璃棒搅拌混合物，然后将其静置一段时间；b. 在分液漏斗中加入少量的饱和氯化钠溶液或固体氯化钠，搅拌并放置一段时间。

the funnel, stir it and let it stand for several minutes.

⑤ Clean the separatory funnel when the extraction is complete and regrease the ground-glass stopcock to prevent sticking and freezing later[⑩].

- **Questions**

① What are the features of a good extraction solvent?

② What is salting out[⑪]? Explain why salting out can improve the efficiency of a liquid-liquid extraction.

③ What will affect the efficiency of a liquid-liquid extraction?

④ When an organic compound is extracted from an aqueous solution with diethyl(10 mL×2), a student performed the operation in accordance with the following procedures: to remove the lower layer after the first extraction and add the second 10 mL of diethyl ether directly to the upper layer remaining in the separatory funnel. After shaking and venting the funnel for several times, only one phase with no interface will be observed. Explain the reason. What should be done next?

2.9 Recrystallization

- **Principle**

The solid product obtained directly from an organic reaction is often impure, containing some by-products and unreacted starting materials. It needs to be further purified before being used. Recrystallization is the most commonly used method to purify a solid substance from a mixture in organic laboratory, also in chemical industry.

The solubility of a solid in a solvent generally varies with temperature. Generally, organic solids are usually more soluble in hot solvent than in a comparable volume of cold solvent. Recrystallization refers to the process that is to dissolve a given amount of mixture of solids in a minimum amount of hot solvent to form a saturated solution, then cool the solution and precipitate the crystals mainly with one component[⑫]. This process is based on the different solubilities of the compound to be purified and the impurities in a certain solvent. The completely insoluble impurity in the solvent even at high temperature can be removed by hot filtration, while the extremely soluble impurity at low temperature will remain in the solution after the crystallization of the pure solids[⑬].

The recrystallization procedures involve the following six steps[⑭]:

---

⑩ 萃取结束后，应立即清洗分离漏斗，并在磨口活塞处涂上凡士林，以防止事后粘连或冻结。

⑪ 盐析。

⑫ 重结晶是指将一定量的固体混合物溶解在热的溶剂中，制成热的饱和溶液。然后冷却此溶液，以析出纯净固体物质的过程。

⑬ 重结晶是利用溶剂对被提纯物质和杂质溶解度的不同，使杂质通过热过滤除去或冷却后被留在母液中，从而达到分离提纯的目的。

⑭ 重结晶的一般过程分为六步：①选择合适的溶剂；②溶解粗品，制备热的饱和溶液；③活性炭脱色；④趁热过滤，除去活性炭和其他不溶性杂质；⑤冷却结晶；⑥抽滤洗涤。

① To select a suitable solvent.

② To dissolve the compound to be purified at the boiling point of the selected solvent to form a saturated solution.

③ To decolorize the saturated solution with activated carbon.

④ To remove the activated carbon and insoluble impurities by hot filtration.

⑤ To cool the solution and precipitate the pure crystalline solid.

⑥ To filter the cooled solution and collect the purified solid and dry the crystal.

Each step of the sequence will be discussed in the following subsections:

1. Solvents Choosing

The choice of a suitable solvent is perhaps the most critical step in the recrystallization process. As an ideal recrystallization solvent, the following criteria must be met[16]:

① The solvent should not react with the substance to be purified.

② The solvent can dissolve the compound to be purified at its boiling point, while dissolving very little or even none at room temperature.

③ The impurities are either insoluble in the hot solvent or soluble in the cold solvent.

④ The solvent should possess a boiling point below the melting point of the compound to be purified.

⑤ The solvent can give out trim crystals of the purified compound at low temperature.

⑥ The solvent should have a relatively low boiling point (60-100 ℃).

⑦ The solvent should be cheap, easily available, and nontoxic.

The most commonly used solvents for recrystallization and their physical properties are listed in Table 2.2.

Table 2.2 Some commonly used solvents in recrystallization

Solvent	Polarity	Miscibility in water	Boiling point(760 mmHg)/℃
water	very high	N/A	65
methanol	high	+	65
ethanol	high	+	75
acetone	intermediate	+	60
ethyl acetate	low	−	77
dichloromethane	low	−	40
toluene	nonpolar	−	110
petroleum ether	nonpolar	−	30-90
cyclohexane	nonpolar	−	81
hexane	nonpolar	−	69

The choice of a suitable solvent is based on the principle "like dissolves like". Generally, polar substances are normally soluble in polar solvents, whereas nonpolar ones are more soluble in nonpolar solvents. The chemical literature offers the desired information about solvents

---

[16] 合适的溶剂必须具备以下条件：①溶剂不与被提纯物质发生反应；②溶剂在较高的温度（沸点附近）能溶解被提纯物质，而在室温只能溶解少量或不溶；③对杂质的溶解度非常大或非常小；④溶剂的沸点要低于被提纯物质的熔点；⑤能得到较好的结晶；⑥溶剂的沸点不要太高，也不宜太低；⑦溶剂的价格低，毒性小，易回收，操作安全。

that may be used for the recrystallization of known compounds. For the recrystallization of those new organic compounds, the appropriate solvents can only be chosen from the solubility test with the following procedures:

Place about 0.1 g of the compound to be recrystallized in a test tube, add 1 mL of the solvent and shake the mixture. The solvent is considered unsuitable if the compound dissolves immediately in a cold or gently warmed solvent. If no solubility is observed, heat the solvent to boil while shaking. If the compound does not dissolve in the boiling solvent, add more solvent with an amount of 0.5 mL one time(heat the mixture to boil while shaking after each addition)until it is completely dissolved. If this compound does not dissolve when the total amount of solvent reaches 4 mL, the solvent will be considered not suitable. If it can dissolve in 1-4 mL of the hot solvent and precipitate crystals after the solution is cooled down, this solvent may be suitable. This procedure should be repeated with other solvents until an appropriate solvent is identified with the highest yield and best crystal structure of the compound.

If a single solvent fails to satisfy the requirement, solvent pairs(the two solvents must be miscible)are often picked for recrystallization. The compound to be recrystallized should be quite soluble in one solvent of the pairs and comparatively insoluble in the other[17]. There are several pairs of solvents often used for recrystallization, such as water-ethanol, ethanol-ethyl ether, ethanol-acetone and benzene-petroleum, etc.

2. Solid Dissolution

The amount of the solvent to be used is also critical in the crystallization process, which not only depends on the amount of the substance to be recrystallized, but is also affected by the solubility of the substance in the chosen solvent at different temperatures. The required amount of the solvent can be calculated according to the solubility of the substance in the solvent to be used at its boiling point. Then a certain amount of solvent less than the calculated amount will be first added to the substance and the mixture will be heated to boil. Then, add the leftover solvent slowly bit by bit while reheating the mixture to boil after each addition until all the substances to be recrystallized completely dissolve in the boiling solution. Additional 10%-20% of the total amount of solvent is usually recommended to be added to prevent premature crystallization caused by the decrease of the temperature during hot filtration[18].

Most of the recrystallization solvents are highly flammable and volatile. In this case, it is prohibited to heat them directly with an open flame. Beakers are not suitable vessels for recrystallization, a flask(round-bottom flask or Erlenmeyer flask)attached with a reflux condenser is therefore often used for recrystallization. Additionally, zeolites should be added to the recrystallization mixture to avoid solution bumping while boiling.

---

[17] 混合溶剂就是将对被提纯物质溶解度很大的和溶解度很小的而又相容的两种溶剂混合而得到的溶剂。

[18] 为了防止在热过滤过程中因温度下降而导致的过早结晶现象，通常要再额外多补加溶剂总量的10%～20%。

3. Decoloration and Hot Filtration[19]

If the hot solution of the desired substance to be dissolved is colored, the decoloration step is necessary for the recrystallization. Activated carbon is always employed to remove the colored impurities. Some activated carbon will be added after the solution is slightly cooled, and the mixture will be reheated to boil for 3-5 min. Then, the impurities-absorbed activated carbon as well as the insoluble impurities will be removed by vacuum filtration when the solution is hot(hot filtration). The Büchner funnel and filter flask used for hot filtration should be pre-heated by soaking in hot water to minimize premature crystallization in the funnel and the flask.

4. Crystallization

After hot filtration, the hot filtrate is allowed to be cooled to room temperature. Some crystals will precipitate slowly in the container as the temperature of the filtrate decreases. If no crystals appear after cooling for a long time, a few crystal seeds can be added into the filtrate or the inside surface of the container should be rubbed vigorously with a glass rod to initiate the crystallization[20].

5. Crystal Collection

Upon the completion of the crystallization, transfer the mixture to a Büchner funnel bearing a filter paper and collect the recrystallized crystals by vacuum filtration. If there are still some crystals remaining on the wall of the container after transferring, rinse them with the filtrate and transfer them to the Büchner funnel again. Wash the crystals on the Büchner funnel with a small amount of pure cold solvent to remove any supernatant liquid adhering to the crystals. Then, compact the crystals with a spatula when the vacuum is applied to remove as much solvent as possible.

6. Crystal Drying

The crystals collected by vacuum filtration are still wet and and should be further dried. The removal of the last traces of the solvent from the crystals may be accomplished by air-drying or drying under vacuum. The latter is recommended because it can make the drying process completed more quickly.

- **Apparatus for Recrystallization**

Typical apparatus for recrystallization is shown in Figure 2.9.

The important pieces include: Erlenmeyer flask, reflux condenser, graduated cylinder, Büchner funnel, filter flask and water aspirator.

---

⑲ 脱色和热过滤。
⑳ 若滤液冷却一段时间后仍无晶体析出，可以通过加入晶种或用玻璃棒摩擦容器内壁来促进结晶。

• **Procedures for Recrystallization**

① Place the substance to be recrystallized as well as several zeolites in the Erlenmeyer flask. Then, add the chosen solvent to the flask and heat the mixture to boil.

② Add more solvent in small portions until all the solids are dissolved in the boiling solution if there are some solids that are still not completely dissolved. Then, add 10%-20% more of the total amount of solvent to the solution and reheat the solution to boil.

Figure 2.9　Apparatus for recrystallization

③ Cool the solution slightly, add about 0.1 g of activated carbon⑫ to it, and then heat the mixture to boil for 3-5 min. Do not add activated carbon to the boiling solution in case of violent bumping⑬.

④ Filter the hot solution as soon as possible and transfer the hot filtrate into a clean container.

⑤ Allow the filtrate to be cooled to room temperature until the crystallization is completed.

⑥ Remove the solvent by vacuum filtration and wash the crystals with a little amount of cold solvent for 1-2 times.

⑦ Collect the crystals and dry them in the air.

• **Tips for Recrystallization**

① Use a flask equipped with a condenser for recrystallization, not a beaker.

② Remember to heat the solution to boil after adding the solvent to each portion in the step of solid dissolution. Always use the minimum amount of hot solvent to dissolve the solids⑭.

③ Make accurate experimental observations to determine whether the compound to be purified contains an insoluble impurity. If no further dissolution of a solid is observed when 1-2 mL of solvent is added to the mixture, the solid should be insoluble impurity⑮.

④ Don't add too much activated carbon to avoid the adsorption of the desired substance.

⑤ Remember to pre-heat the Büchner funnel and filter flask before hot filtration.

⑥ Don't cool the flask containing the hot filtrate rapidly by immersing the container in water or an ice-water bath, which will lead to the formation of very small crystals that may adsorb impurities from the solution⑯.

⑦ Use the minimum amount of cold solvent to wash the crystals. Pay attention to the accurate operation procedure for washing a solid.

---

⑫ 活性炭的用量一般为固体质量的1%～5%。

⑬ 不要向正在沸腾的溶液中直接加入活性炭，以免溶液暴沸。

⑭ 在溶解固体时，每加一次溶剂都要把溶液加热至沸腾后再观察固体的溶解情况。尽量用最少量的热溶剂来溶解被提纯的物质。

⑮ 如果向沸腾的溶液中继续加入1～2 mL溶剂后，并没有观察到溶液中的固体物质进一步溶解，则该固体应为不溶性杂质。

⑯ 不要用冷水浴或冰水浴快速冷却滤液，这会导致形成的晶体不仅非常小，而且还可能包夹着溶液中的杂质。

⑧ Don't dry the solids recrystallized from volatile organic solvents in an oven[15].

• **Questions**

① List the main steps for the recrystallization and briefly explain the purpose of each step.

② Why should the amount of the solvent to be used in recrystallization be neither too much nor too little?

③ When performing a hot filtration, what might happen if the Büchner funnel and the filter flask are not preheated in hot water before filtration?

④ What problems might occur if crystallization precedes too rapidly?

⑤ Why sometimes does crystallization not occur even after the filtrate in an ice-water bath is cooled down? What steps can be taken to initiate crystallization?

⑥ The goal of the recrystallization procedure is to obtain purified material with a maximized recovery. For each of the listed item, which would be adversely affected? Explain the reasons.

a. In the dissolution step, an unnecessarily large volume of solvent is used.

b. The crystals obtained after filtration are not washed with the fresh cold solvent before drying.

c. The crystals obtained after filtration are washed with fresh the hot solvent.

d. A large quantity of decolorizing carbon is used.

e. Crystallization is accelerated by immediately placing the flask of hot solution in an ice-water bath.

2.10 Sublimation

• **Principle**

All substances have three phases, namely gas phase, liquid phase and solid phase. As shown in the phase diagram[16] (Figure 2.10), the phase of a substance is related to the external pressure and temperature applied to the substance. Each curve in the diagram demonstrates the equilibrium between two phases. The cross point of the three curves is called triple phase point, where the three phases exist together in equilibrium[17]. A solid with a sufficiently high vapor pressure below its triple phase point may be directly converted into gas without going through the intermediate liquid state. The best known example is the conversion of the solid form of carbon dioxide (dry ice) directly into CO_2 gas at atmospheric pres-

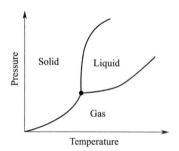

Figure 2.10 A typical phase diagram

---

⑮ 千万不要用烘箱干燥从挥发性有机溶剂中结晶出来的固体。

⑯ 相图。

⑰ 三条曲线的交叉点称为三相点，三相在此处平衡存在。

sure. Sublimation is the complete process of evaporation from the solid phase and condensation from the gas phase to form pure crystals without passing through the liquid phase[159].

Sublimation, like recrystallization, is also a method to purify organic solids, which is commonly used to separate a sublimable substance from insublimable impurities. Sublimable substances are relatively nonpolar and fairly symmetrical, and usually have low molecular weights and weak intermolecular attractive forces, such as naphthalene, 1,4-dichlorobenzene and ferrocene. In general, only organic substances with the following properties are suitable to be purified by sublimation[160]:

① Organic substances with sufficiently high vapor pressure(>266.6 Pa)below their triple phase points.

② Organic substances thermostable enough to vaporize without decomposition or melting.

③ Organic substances with vapors to be easily condensed to form crystals.

④ Organic substances containing impurities that are not easy to sublime.

Sublimation occurs when the vapor pressure of the solid equals the applied external pressure. In this case, if the substance has less vapor pressure at atmospheric pressure or decomposes easily when heated, its sublimation is more appreciably to be carried out at reduced pressure. Vacuum sublimation is more popular in organic laboratory because it makes the decomposition and melting of the solids less likely to occur during the sublimation process[161].

- **Apparatus for Vacuum Sublimation**

A sublimation apparatus includes a vessel for the evaporation of the substance to be purified and a cold finger surface to condense the vapor to a pure sublimed product. The important pieces include: test tube, filter flask, filter adapter, vacuum source and appropriate heat source. The typical apparatus for vacuum sublimation of small quantities is shown in Figure 2.11.

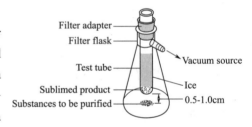

Figure 2.11 Apparatus for vacuum sublimation

- **Procedures for Vacuum Sublimation**

1. Place the sample to be purified in a filter flask.

2. Fix the inner test tube through the filter adapter and adjust the position of the test tube so that it is 0.5-1.0 cm above the bottom of the filter flask. Then, fill the inner test tube with ice.

3. Turn on the vacuum source and heat the sample gently just below its melting point[162].

---

⑲ 物质不经过液态而直接变成气态的相变过程称为升华。

⑳ 可用升华来提纯的物质必须具备以下特点：①在其三相点下具有较高的蒸气压；②具有足够的热稳定性，气化时不会分解或熔化；③其蒸气易冷凝成固体；④含有的杂质不容易升华。

㉑ 减压升华特别适用于在常压下蒸气压不太大或加热易分解的物质。减压升华在有机实验室中应用十分广泛，因为这样可以避免升华过程中固体的分解和熔化。

㉒ 升华过程加热要缓慢，并控制加热温度低于被提纯物质的熔点。

4. Stop heating and remove the heat source when all the volatile substances have been sublimed and only the nonvolatile impurities remain at the bottom of flask.

5. Disconnect the rubber tubing from the vacuum source slowly, and then turn off the vacuum source.

6. Take out the inner test tube carefully and collect the purified solid using a spatula.

- **Tips for Vacuum Sublimation**

1. Control the heating rate and temperature, so as to avoid melting or decomposition of the sample being sublimated.

2. Add additional ice if all the ice in the inner test melts to ensure the cooling effect.

3. Stop the sublimation operation temporarily if the inner test tube is overloaded with the sublimates. After removing the sublimates from the test tube, reassemble the apparatus and resume the heating.

- **Questions**

1. What criteria should be used to determine whether recrystallization or sublimation will be selected for the purification of an organic solid?

2. An organic solid has a vapor pressure of 900 torr at its melting point (100 ℃). Explain how you will purify this compound.

3. Why is sublimation usually carried out at reduced pressure?

2.11 Measurement of the Melting Point

- **Principle**

The melting point of a solid substance refers to the temperature at which solid and liquid phases exist in equilibrium with one another[15]. If heating a mixture of the solid and liquid phases of a pure solid substance at its melting point, no change of the temperature will be observed until all the solids have melted into liquids. The melting point of a pure solid substance actually denotes a melting range, which is usually recorded from the temperature at which the first drop of liquid appears to the temperature at which the solid melts completely into a clear liquid[16]. The melting ranges reported for pure substance will normally be no more than 1 ℃. Meanwhile, the existence of a small amount of impurities will lower the melting point and enlarge the melting range of a substance. In general, the more impurities a substance contains, the lower its melting point and the larger its melting range will become. Thus, the difference between the melting point and its range of a substance reported in literature with observed in experiment can give us a roughly approximate purity of the substance obtained from the experiment.

In summary, melting point measurement provides us a convenient and rapid method to

---

[15] 熔点是指一个大气压下固体化合物固相与液相平衡时的温度。

[16] 纯净的固体物质的熔点实际上是一个范围（熔程），指从第一滴小液滴出现的温度开始，到固体恰好完全转化为液体的温度结束。

identify and ascertain the purity of a solid substance. The substance with a higher melting point and a narrower melting range is ordinarily pure[16].

• **Measurement Methods and Apparatus of the Melting Point**

There are many methods available for the measurement of the melting points of organic substances, among which, the capillary melting-point method and melting-point microscopic method are commonly used in organic laboratory[17]. These two methods will be separately discussed in the following parts.

1. Capillary Melting Point Method

(1) Sample preparation

The subdivision, dryness and the amount of the substance will affect the accuracy of melting point measurement. The substance to be measured should be dried completely and crushed into a fine powder before the measurement.

Place a small amount of the substance to be measured on a clean watch glass and tap the open end of the capillary tube in the powdered substance. A small amount of substance will stick in the open end of the tube. Then, invert the capillary tube carefully and drop it down a long glass tubing with the bottom on the lab bench. This procedure will be repeated several times until the substance in the capillary tube has packed tightly at the bottom with 2-3 mm in height[18].

(2) Heating source

The heating source for the capillary melting-point method is provided by a heating fluid filled in a Thiele tube(Figure 2.12). The fluid is heated at the side-arm of the Thiele tube by a burner. With the convection currents of the fluid, the heat will be distributed to all parts of the tube, which allows the sample in the capillary tube to be heated evenly during the measurement. The heating rate can cause the observed melting point and its range to be different from the results reported in the literature. The most accurate result will be obtained at the heating rate of temperature rising about 1-2 ℃ per minute when the temperature is about 10℃ below the expected melting point[19].

(3) Apparatus for the capillary melting-point method

Typical apparatus for the capillary melting-point method is shown in Figure 2.12.

The important pieces include: heating source, closed-end capillary, thermometer, heating fluid, Thiele tube, cork, etc.

---

[16] 综上所述，熔点测量为我们识别和确定固体物质纯度提供了一种方便快捷的方法。熔点高、熔点范围窄的物质通常是纯的。

[17] 有机实验室最常用的熔点测定的方法是毛细管法和显微熔点仪测定法。

[18] 毛细管中样品的填充要均匀且紧密，样品的高度约为 2～3 mm。

[19] 加热的速度会影响到熔点测定的准确性。当温度到达比标准熔点低约 10℃时，要控制加热的速度为每分钟升温 1～2 ℃。

(4) Procedures for capillary melting-point method

① Put the heating fluid into the Thiele tube and make sure that the liquid level is flushed with the upper side of the side-arm of the Thiele tube[169].

② Clamp the Thiele tube to an iron stand.

③ Place the sample to be measured into a capillary tube, and tie the capillary tube with the thermometer together using a rubber band and make them attached as close as possible.

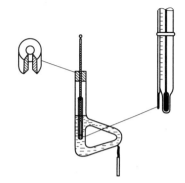

Figure 2.12 Typical apparatus for capillary melting-point method

④ Put the capillary tube and the thermometer into the Thiele tube and let them immersed in the heating fluid. Make sure that neither the thermometer nor the capillary tube touches the glass wall of the Thiele tube, and that the mercury bulb of the thermometer is in the middle of the side-arm of the Thiele tube[170].

⑤ Heat the Thiele tube and adjust the heating rate so that the temperature rises at a moderate rate. In the case of a temperature about 10 ℃ below the expected melting point of the sample, reduce the heating rate and keep the temperature rising only 1-2 ℃ per minute.

⑥ Observe the melting process and write down the melting range of the sample.

⑦ Turn off the flame, remove the used capillary tube and discard it into the glass disposal container.

Note: For a new compound whose melting point is not yet reported, one can heat the sample at a relatively fast rate to get a rough melting point, and then follow the procedures described above to measure the melting point more accurately[171].

2. Melting-point Microscopic Method

The common advantages of the melting-point microscopic method include: ①The amount of sample to be measured is small; ②The melting changes of substances can be observed visually during the heating process; ③The melting point can be measured in a wide range, from room temperature to a temperature as high as 300 ℃[172]. There are many types of micro melting point apparatus. Figure 2.13 shows the X-6 micro melting point apparatus.

Figure 2.13 X-6 micro melting point apparatus

---

⑯⁹ 浴液的液面应与提勒管侧管的上端齐平。

⑰⁰ 温度计和毛细管都不接触提勒管的内壁,温度计的水银球处于提勒管的两侧管中间。

⑰¹ 测定尚未见文献报道的新化合物的熔点时,可以以较快的速度加热样品,得到粗测的熔点,然后再按照上述步骤精测熔点。

⑰² 显微熔点测定法的优点在于:①药品的用量少;②可以直观地观看到化合物熔化的过程;③熔点测定的范围广,能测量熔点在室温至 300 ℃范围的样品。

The general procedures for the melting-point microscopic method are summarized as follows:

① Place the sample between two clean and dry glass plates and put the plates on the heating plate.

② Adjust the objective lens and eye lens of the microscope until all the samples can be seen clearly.

③ Turn on the heater and heat the sample at a moderate rate.

④ Reduce the heating rate when the temperature is about 10 ℃ below the expected melting point of the sample and then keep the temperature rising at a rate of only 1-2 ℃ per minute.

⑤ Observe the melting process and record the melting range.

⑥ Stop heating and clean the glass plates with acetone.

• **Tips for Melting Point Measurement**

① The capillary tube or the two glass plates filled with the sample to be measured must be dry and clean.

② The sample in the capillary tube should be finely ground and packed tightly, so as to ensure the rapid and uniform conduction of the heat.

③ The thermometer must be calibrated before performing a melting point determination.

④ The heating rate must be controlled. The heating rate can be rapid at first and then must be reduced gradually when the temperature reaches close to the melting point.

⑤ If a second measurement is required, the apparatus should be cooled down and the temperature should be kept at least 20 ℃ below the expected melting point before the next measurement⑰.

• **Questions**

1. Describe the errors that may cause the observed melting point of a pure compound:
① to be lower than the correct melting point;
② to be higher than the correct melting point;
③ to be broad in range(over several degrees).

2. Do the present impurities always lower the melting point of a solid substance? Explain your answer.

3. Both benzoic acid and 2-naphthol melt at about 122 ℃. In the case of an unknown compound posses a melting point also around 122 ℃, please design a procedure to determine whether this compound is benzoic acid or 2-naphthol.

2.12 Refluxing

• **Principle**

Many organic reactions occur very slowly at room temperature, heating is necessary for

---

⑰ 进行第二次熔点测量时，把加热温度降至低于待测样品熔点 20 ℃以下再进行测量。

most organic reactions. In general, the reaction rate will be doubled for every 10 ℃ rising in the reaction temperature. If heating is carried out in an open system, the vapor of the solvent and other liquid materials tend to escape to the atmosphere, thereby causing potential safety hazards and the change of reactant concentration. On the contrary, if the heating system is closed, pressures in the reaction flask will be built up with the increasing temperature, which may lead to an explosion. Refluxing refers to the process that a reaction mixture is heated to boil in a flask, and the volatile substances in the mixture are continuously condensed to liquids and dropped back to the flask. This is a simple method that allows reactions to be conducted at higher temperatures without a loss of any reagents⑭.

Refluxing is performed by means of a condenser mounted vertically above the reaction flask. When the reaction mixture is heated to boil, the volatile reagents will be vaporized and escape from the flask. The rising vapor is then condensed to liquid by a cold condenser and flows back to the flask guided by the condenser. There are two common types of condensers available for refluxing: one is the Allihn condenser, while the other is the air condenser. Generally, an Allihn condenser with water flowing through its outer jacket is usually used in the case of a boiling point of reaction mixtures lower than 140 ℃. If the boiling point of reaction mixtures exceeds 140 ℃, an air cooling is enough to condense the vapor and is therefore recommended to avoid bursting of the Allihn condenser caused by excessive temperature differentials between the tap water and the rising vapor⑮.

The heating rate should be considered for the performing of refluxing. If the heating rate is too rapid and exceeds the maximum cooling capacity of the condenser, the volatile reagents in the mixture may also escape from the top of the condenser. It is always preferred to control the heating rate at a moderate level to make sure that the vapor level does not exceed the 1/3 height of the condenser, and that the vapor-liquid conversion level is at the half of the first ball of the Allihn condenser⑯.

- **Apparatus for Refluxing**

Four common reflux apparatus are shown in Figure 2.14, where the Figure 2.14(a) depicts the general reflux apparatus. If the reaction mixture is water-sensitive, a drying tube containing anhydrous calcium chloride will be connected to the top of the condenser, as shown in Figure 2.14(b). Figure 2.14(c) presents the reflux apparatus with a gas-trap system, which is applied for the absorption of poisonous gases generated during reactions. Figure 2.14(d) shows a typical apparatus that allows liquid reagents to be added to the reaction flask while the reaction is refluxing⑰.

---

⑭ 回流是指将反应混合物在烧瓶中加热至沸腾，混合物中的挥发性物质遇冷不断凝结成液体并回到烧瓶中的过程。回流使反应可以在比较高的温度下进行，且不会造成试剂的损失。

⑮ 如果反应混合物的沸点超过140 ℃，空气足以将蒸气冷凝，此时应使用空气冷凝管，以避免因冷凝水和上升的蒸气间温差过大而造成球形冷凝管爆裂。

⑯ 回流时，要控制加热的速度，使蒸气的水平高度不超过冷凝管的1/3，且蒸气浸润处不超过球形冷凝管第一个球的一半。

⑰ 图2.14(d) 所示为滴加液体试剂和加热回流可同时进行的实验装置。

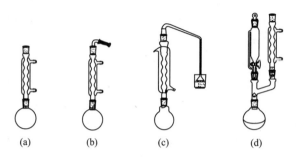

Figure 2.14　Common apparatus for refluxing

The important pieces for refluxing include: round-bottom flask, zeolites, condenser, drying tube, gas-trap system and addition funnel with pressure equilibrating arm.

• **Procedures for Refluxing**

1. Place the reaction mixtures and zeolites into a flask and clamp the flask to an iron support.

2. Attach rubber tubes to a Allihn condenser and connect the condenser to the flask.

3. Introduce tap water into the condenser, and then heat the flask with a heating mantle.

4. Control the heating rate and reflux the reaction mixture for the designated time.

5. Turn off the heating mantle, so as to allow the reaction mixtures to be cooled down to room temperature.

6. Close the tap water and disassemble the apparatus.

• **Tips for Refluxing**

1. The volume of the reaction mixtures to be refluxed should not exceed two thirds that of the flask[18].

2. Be sure to introduce the tap water to the condenser before heating the reaction mixtures.

3. The tap water should flow into the water jacket of the condenser at the bottom inlet and out at the top outlet[19].

4. The water in the outer jacket of the condenser should not flow too fast to avoid the rubber hose popping off the condenser.

• **Questions**

1. Is it proper operation to add zeolites down through the condenser to a boiled solution?

2. What should we do if we realize that the zeolites are not added at the time when the solution has started to boil?

3. Why the cooling water must be introduced to the water jacket of the condenser at the

---

[18]　回流时，烧瓶内液体的体积不要超过烧瓶体积的 2/3。
[19]　冷凝水应从冷凝管的下口进，上口出。

bottom inlet and out at the top outlet?

4. Why an air condenser must be used for refluxing when the boiling points of mixtures are higher than 140 ℃?

5. A certain reaction requires an hour to reach complete refluxing in diethyl ether(b. p. = 35 ℃). How long does the refluxing procedure require to complete the reaction when this reaction is carried out in tetrahydrofuran(b. p. = 65 ℃)?

2.13 Stirring Methods

• **Principle**

If a reaction is heterogeneous or involves dropping of reactants, stirring must be applied to the reaction mixtures, which ensures the complete interaction among the reactants and thus facilitates the reactions. Meanwhile, it makes the concentration and homogeneous temperature of the reactants in the flask more uniform, thereby reducing the occurrence of side reactions. In addition, stirring can also be used to replace the zeolites to maintain smooth boiling and avoid bumping for a boiling mixture[18]. Generally, there are several ways to stir the reaction mixtures effectively, such as hand stirring, magnetic or mechanical stirring.

1. Hand Stirring

The simplest stirring approach is hand stirring. When a reaction is carried out at room or moderate temperature, a glass rod can usually be used to manually mix the reactants in the flask as long as the reactants are least volatile and insensitive to air and moisture. For a refluxing reaction, it is possible to mix the reactants in the flask evenly just by picking up the iron support and periodically moving the entire assembly in a circular motion.

2. Magnetic Stirring

Magnetic stirring is the most common technique for mixing the contents in a flask in organic laboratory. Magnetic stirring equipment is made of a stir bar and a magnetic stirrer, consisting of a large bar magnet and a variable-speed motor. Most of the stir bars are magnets coated with a chemically inert and nonflammable substance, such as polytetrafluoroethylene(teflon). Stir bars are made in several sizes with different shapes to match the different shapes and opening sizes of the reaction vessel, as well as the different volumes of the reactants to be stirred.

A stir bar is normally put into the reaction flask before any other reactants, and it should gently slide down into the tilted flask instead of being thrown directly, so as not to crack or

---

⑱ 搅拌不仅能使反应物之间充分接触，更有利于反应的进行，缩短反应时间；搅拌也会使反应的温度和浓度更均匀，避免副反应的发生。同时，搅拌还可以替代沸石，使加热反应得以平稳进行。

break the bottom of the flask[18]. The flask containing a stir bar should be placed in the middle of the stir plate and be adjusted to be close to the magnet of the stir plate so that the stir bar can rotate smoothly without wobbling. The stirring rate may be adjusted using the control dial on the motor, and increased slowly to reach the optimum rate. The stirring rate should not be too high, so as to minimize splashing. Magnetic stirring is usually combined with heating in organic laboratory, which makes simultaneous heating and stirring of reaction mixtures possible.

3. Mechanical Stirring

The power of magnetic stirring is normally too limited to stir a high viscous system or a heterogeneous mixture. In this case, mechanical stirring will be employed. Mechanical stirring device consists of a variable-speed electric motor and a stirring shaft, usually made of an inert material, such as stainless steel, teflon or glass. One end of the stirring shaft is connected to the electric motor with a short length of rubber tubing, while the other end is inserted into the reaction flask with 0.5 cm distance between the

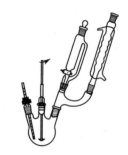

Figure 2.15　Apparatus for mechanical stirrer

shaft and the bottom of the flask. It is important to select an appropriate stirring shaft based on the reaction medium and the different sizes of the reaction flask to ensure the stirring shaft not to touch the inner wall of the flask or the thermometer installed in the flask[19]. It is critical to check the integrity and immobility of the apparatus, and make sure that all the clamps and the joints are not loose while the motor is running. The stirring rate could be controlled by adjusting the transformer power, and carefully increasing the power to achieve the best stirring.

Other operations can also be performed by mechanical stirring. For example, the following apparatus (Figure 2.15) can be used when the dropwise addition of reaction reagents is performed, meanwhile the reaction mixtures are being stirred and heated.

2.14　Water-free and Air-free Operation

When reactants are sensitive to moisture or air, a water-free and air-free environment should be applied to the reaction system. Less strict water-free and air-free condition can be achieved in the undergraduate organic laboratory by the following simple methods: ①to thoroughly dry all the glassware and reagents before conducting the reaction; ②to assemble a drying tube filled with drying agents to the reaction apparatus and prevent moisture from ente-

---

[18] 向反应瓶中加搅拌子时，应先将反应瓶倾斜，然后让搅拌子慢慢地滑入瓶底。不要将搅拌子竖直扔进反应瓶，以避免打碎反应瓶的底部。

[19] 机械搅拌时，搅拌桨的高度应离反应器的底部约 0.5 cm 的距离。要根据反应的介质和反应容器的大小选择合适尺寸的搅拌桨，确保搅拌过程中，搅拌桨不会碰到反应器的内壁和在反应器内安装的温度计。

ring the reaction container; ③to expel the air from the reaction container by using low boiling point volatile solvents, such as diethyl ether, as the reaction solvent; and ④to provide an inert atmosphere to protect the reactions from the air by using a balloon filled with nitrogen gas.

Some reagents, such as organometallic compounds, organoboranes and metal hydrides, can react quickly and vigorously with oxygen, also with moisture in the air. Reactions involving these reagents must be carried out in a strict inert atmosphere with air and water excluded completely[133]. Water-free and air-free operation involves using anhydrous solvents/reagents and dry apparatus, emptying the air and water in the reaction containers, and performing the reactions under inert gas atmosphere. Several special apparatuses including vacuum-line, Schlenk line and glove-box are all commonly used for the creation of water-free and air-free conditions[134]. Vacuum-line and glove-box are rarely used in undergraduate organic laboratory. Herein, the principle and operation of Schlenk line are hereby detailed.

Schlenk line is widely used in organic chemistry laboratory. Some general basic operations, such as reflux, stirring, distillation, addition and filtration, can all be performed under water-free and air-free conditions with the help of Schlenk line. Figure 2.16 depicts an illustrative scheme of the dual manifold Schlenk line, the main components of which are two parallel and interlinked glass tubes, usually known as the manifold in Schlenk line operation. One manifold is connected to the cold trap and vacuum pump, while the other to the inert gas (nitrogen or argon) source[135].

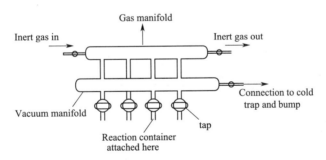

Figure 2.16　An illustrative scheme of dual manifold Schlenk line

There are special triple-valve taps at the junctions of the two manifolds, which are connected to the two manifolds and the reaction container, respectively (Figure 2.17). One can alternate the evacuation and the inert gas refilling process by simply rotating triple-valve taps to provide the reaction system with an inert atmosphere[136].

---

　　❸　一些试剂，如有机金属化合物、有机硼烷和金属氢化物，可以与氧气以及空气中的水分迅速而强烈地反应。因而，涉及这些试剂的反应必须严格在无水、无氧条件下进行。
　　❹　真空线、史兰克系统和手套箱等都是常用的无水无氧反应装置。
　　❺　史兰克系统的主要部件是两根相互平行又彼此相通的玻璃管，通常称为双排管。其中，一排管与冷阱和真空泵相连，另一排管连接到惰性气体（氮气或氩气）源。
　　❻　在双排管的连接处有一个特殊的三通开关，它分别与双排管和反应器相连。通过简单地旋转三通开关可以交替进行抽真空和充氮气的操作，从而为反应体系提供惰性氛围。

There are four to six taps of a Schlenk line typically, making it possible to carry out several reactions and operations simultaneously. It is crucial to operate Schlenk line correctly so as to provide a satisfied water-free and air-free reaction condition. The general operation procedures for Schlenk line are as follows:

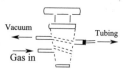

Figure 2.17　A triple-valve tap in Schlenk line

1. Dry the cleaned apparatus used in the experiment in an oven, and then cool them in a desiccator before use. The reagents and the solvents to be used must be water-free.

2. Connect the reaction container to the Schlenk line using a rubber tubing.

3. Repeat the process of alternating the evacuation and inert gas refilling for three times to remove the trace of water and oxygen in the reaction container[⑰].

4. Add the reagents and solvents to the reaction container under inert gas. The solid reagents can be added before the first vacuumizing. The solvents and the liquid reagents should be added after the reaction container being purged by three times co-operation of vacuumizing and gas-refilling[⑱]. Liquid substances are usually added to the container using a transfer syringe with a long flexible needle or a cannula, a long double-ended needle.

5. Maintain a constant inert gas pressure in the Schlenk line until the reaction is completed. Pay attention to the bubbler connected to the inert gas resource and control the bubbling rate to avoid the waste of inert gas.

6. Close the inert gas cylinder timely after the reaction is completed.

The process of water-sensitive and air-sensitive operation is usually continuous. Each step should be performed carefully in case of abandoning the experiment midway[⑲].

2.15　Chromatography Techniques

Introduction

Chromatography is one of the most effective techniques to separate and purify organic compounds from complex mixtures. As implied by the name, chromatography was originally used to separate mixtures of colored substances. Afterwards, chemists confirmed the efficiency of using this technique to separate colorless substances by means of visualization techniques. Nowadays, chromatography techniques have been developed into extensively used methods for both qualitative separations and quantitative analysis in the organic laboratory.

Chromatography analysis is based on the principle that the different components of a

---

⑰ 抽真空和充氮气的操作应交替重复三次，以完全除去反应器内痕量的水和氧气。

⑱ 固体试剂可在第一次抽真空前加入，溶剂和液体试剂则应在反应器经过三次抽真空和充氮气的操作后在惰性气体氛围下加入。

⑲ 无水、无氧的实验操作过程通常是连续的。每一步的操作都要认真仔细，并确保操作的无误，以免中途出错而放弃实验。

mixture distribute unequally between two immiscible phases, usually one mobile phase and one stationary phase. The former generally involves a liquid or a gas, and the latter may adopt a solid or a liquid. In chromatography methods, a mixture dissolved in the mobile phase is carried by the mobile phase and moves through the stationary phase. In this process, different components of the mixture demonstrate distinct mobility because they have different interactions with the stationary phases. The component that interacts more strongly with the stationary phase moves more slowly in the direction of the mobile phase. Such a difference in mobility can make the different components of the mixture well separated after the mixture moves far enough with the mobile phase.

According to the principle and operating conditions, chromatography methods can be divided into thin-layer chromatography (TLC), column chromatography, gas chromatography (GC) and high performance liquid chromatography (HPLC), etc[18].

1. Thin-layer Chromatography

Thin-layer chromatography (TLC), a simple, inexpensive and efficient method for rapid analysis of small quantities of samples, is indeed a form of solid-liquid adsorption chromatography. It is often used in organic laboratory to identify the components in a mixture, to separate compounds at a scale between microgram to gram, to monitor the progress of a reaction as well as to be used as a forerunner of column chromatography[19].

- **Principle**

In thin-layer chromatography analysis, a solid adsorbent coated on a glass plate serves as the stationary phase. The most commonly used adsorbents are silica gel and aluminum oxides. The mobile phase involves a pure solvent or a solvent combination. To carry out the thin-layer chromatography analysis, a small amount of the sample to be analyzed is dissolved in an appropriate solvent and then spotted onto the adsorbent near one end of a plate, which is then placed in a closed developing chamber containing enough of the developing solvent (Figure 2.18). As the developing solvent migrates up the plate by capillary action, it will carry the components of the sample to move up on the plate at different rates. The developing rate of the individual component depends upon the different intermolecular interactions between the components with the stationary phase. Upon the completion of the development, a series of separate spots on the plate may be observed, each representing a single component of the sample (Figure 2.19)[20].

Most organic compounds, except those volatile compounds with boiling points below 100 ℃(760 mmHg), can be expediently analyzed by using thin-layer chromatography.

---

[18] 根据作用原理和操作条件的不同，色谱法可分为薄层色谱、柱色谱、气相色谱和高效液相色谱。

[19] 薄层色谱是一种微量、快速、简便的分离分析方法，在有机实验室中常用于反应进程的追踪、化合物的鉴定、少量样品的纯化以及柱色谱的先导等方面。

[20] 当展开剂通过毛细作用在板上展开时，样品中的不同组分会以不同的速度随展开剂在板上向上移动。不同组分的展开速度取决于该组分与固定相之间吸附能力的强弱。展开结束后，板上会出现一系列独立的斑点，每个斑点代表样品中的一个组分。

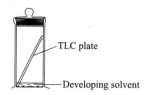

Figure 2.18　Developing chamber containing a thin-layer plate

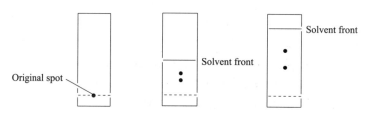

Figure 2.19　Steps in the development of a TLC plate

- **Materials and Methods in Thin-layer Chromatography**

(1) Adsorbents

The most frequently used solid adsorbents for thin-layer chromatography analysis are silica gel and aluminum oxide[⑱]. Silica gel is slightly acidic and is usually used to separate acidic and neutral compounds, while aluminum oxide is available in acidic, basic, and neutral forms and can be adapted to separate nonpolar and polar compounds. Most organic compounds can be bond to these solid absorbents by intermolecular interaction, and the interaction strength varies along with the polarities of different compounds. In general, the more polar the compound is, the more tightly it will be bound to the adsorbents.

The adsorbents coated on the surface of a glass or metal plates are usually used as the stationary phase in thin-layer chromatography. The general procedures for the preparation of TLC plates in laboratory are as follows: to mix the finely ground adsorbent solid with a small quantity of a binder, and then spread the resulting mixture on the plate with the layer with a thickness of 1-2 mm. The thickness of the layer should be as uniform as possible. After being dried at room temperature for a moment, the TLC plates should be activated in an oven for an hour at 110 ℃ to remove any adsorbed water before use[⑲].

(2) Developing solvent

The key to a successful separation is to select a suitable developing solvent with proper polarity. Polar compounds are usually absorbed more tightly to the solid adsorbents and may not move with non-polar developing solvents, while the non-polar compounds, which are bound less tightly to the solid adsorbents, will move too fast with polar developing solvents. Therefore, while carrying out thin-layer chromatography, the selection of an appropriate de-

---

⑱　薄层色谱中最常用的吸附剂是硅胶和氧化铝。

⑲　将铺好的薄层板放置于室温晾干后，需将其放入烘箱内在110 ℃加热活化，以除去吸附剂中的水分。

veloping solvent is mainly based on the polarity of the sample to be analyzed[18]. In general, non-polar developing solvents are often used for the compounds with lower polarity, while polar developing solvents are used for compounds with high polarity.

For a sample featuring unknown polarity, the developing solvent is often selected through a trial-and-error process, which can start with a low polar developing solvent. If the compound in this solvent moves too slowly on a TLC plate, a more polar solvent should be switched so that the compound can move faster, but remember to switch to a less polar solvent if this compound moves too fast. An appropriate solvent for thin-layer chromatography analysis generally gives a R_f value of 0.30-0.60 for a compound.

The commonly used developing solvents are listed from least polar to most polar as: hexane<cyclohexane<toluene<diethyl ether<methylene chloride<ethyl acetate<acetone<ethanol<methanol.

If a single solvent fails to satisfy the requirement, solvent combinations are also used to develop solvents for thin-layer chromatography. The most used solvent combinations are hexane-ethyl ether and dichloromethane-methanol. The polarity of a solvent combination can be controlled by adjusting the different percentages of each solvent in the combination[19].

(3) Sample application

① Dissolution of a sample

The sample to be analyzed should be dissolved in a volatile organic solvent to form a dilute solution with a concentration of 1%-2%. Some anhydrous organic solvents, such as acetone, ethyl acetate and dichloromethane, are commonly used to dissolve the solid samples.

② Spotting on a TLC plate

A base line is drawn on the narrow edges of the plate with about 1 cm away from the bottom, and one end of the capillary tube is dipped into the sample solution to get 2-3 μL of the solution. Then, hold the capillary tube vertically and apply the solution to the TLC plate by touching the capillary tube gently and briefly on the base line. Do not spot the TLC plate with too much sample, in case of large tailing spots and poor separation. The diameter of the drop left on the adsorbent should be less than 2 mm. If the applied spot is too small, repeat the above spotting procedures and apply more samples to the thin-layer chromatography plate at the exactly same place. Do not apply the second time until the first spot is dried[20]. Attention should also be paid to touching the capillary tube to the plate gently to avoid gouging a hole on the adsorbent.

(4) Development of a TLC plate

① Preparation of the developing chamber

Sufficient amount of developing solvent is added into the developing chamber. The sur-

---

⑱ 薄层色谱中展开剂的选择主要是由样品的极性来决定的。

⑲ 如果单一的溶剂达不到理想的分离效果,可以用混合溶剂作为展开剂。混合溶剂的极性可以随着每种溶剂在组合中所占比例的不同而变化。

⑳ 样品点的直径要小于 2 mm。如果样品点过小,可待溶剂挥发后在同一位置重复点样。

face of the developing solvent must be lower than the spots applied on the TLC plate to prevent the sample from being dissolved in the developing solvent, which will ruin the chromatogram[⑱]. Then, cap the developing chamber and place it for 5-10 min to ensure that the chamber atmosphere is saturated with solvent vapors before use[⑲].

② TLC development

Place the sample-loaded thin-layer chromatography plate into the chamber carefully after the plate is dried[⑳]. Recap the developing chamber and allow the developing solvent to rise up the plate through capillary action. Do not lift or disturb the chamber while the TLC plate is being developed. Pick the TLC plate from the developing chamber carefully when the solvent front is 1.0 cm away from the top of the plate. Immediately mark the solvent front with a pencil on the plate and allow the plate to be dried for further analysis.

(5) **Visualization techniques**

Most organic compounds are not colored, and the thin-layer chromatography separation of these compounds cannot be observed directly. The spots on the TLC plate must be visualized using other methods, including:

① Fluorescence

The simplest visualization technique involves the use of adsorbents that contain a fluorescent indicator. A commonly used adsorbent containing a fluorescent indicator in organic laboratory is GF_{254}[㉑], which rarely interferes in any way with the chromatographic results. When a TLC plate containing a fluorescent indicator is irradiated by a short-wavelength ultraviolet lamp(254 nm) in a dark box, a light green fluorescent background will be rendered. Meanwhile, the separated compounds, which can absorb the ultraviolet light, will appear as purple spots on the fluorescent background. Remember to outline each spot with a pencil when the thin-layer chromatography plate is under the UV source to give a permanent record for further chromatographic analysis. This method is only applicable to certain UV reactive substances, such as aromatic compounds.

② Iodine visualization

Place the TLC plate in an enclosed chamber containing some iodine crystals. After several minutes, brown spots, which are formed by interacting of different components in the sample with iodine, will be observed on the TLC plate[㉒]. This process is reversible and the colored spots will disappear after the removal of the thin-layer chromatography plate from the iodine chamber, so spots should also be circled at once.

Besides iodine, there are also some other reagents that will also cause the development of

---

⑱ 展开时,薄层色谱板上样品点的位置必须高于层析缸中展开剂的液面,以防止样品溶解在展开剂中,导致层析失败。

⑲ 将适量的展开剂倒入层析缸中后,要盖上层析缸的盖子并放置5~10min,待层析缸中的空间被展开剂饱和后,方可进行层析。

⑳ 待薄层色谱板上溶剂挥发完后,方可放入层析缸中进行层析。

㉑ 最简单的显色技术是使用含有荧光指示剂的吸附剂。实验室常用的含有荧光指示剂的吸附剂是GF_{254}。

㉒ 碘熏显色法是将晾干后的薄层色谱板放入含有碘的密闭容器中,许多有机物都能和碘形成黄棕色斑点。

a color, such as sulfuric acid, ninhydrin, potassium permanganate, etc. These reagents can react with the individual component to produce colored or dark spots.

(6) Determination of the R_f

The R_f value for each substance is one of the most important data that needs to be recorded in a thin-layer chromatography analysis. R_f (ratio to the front) values are calculated by dividing the distance traveled by a component based on the distance traveled by the solvent[203]. The traveling distance of a component is measured from the origin where the initial spot is applied to the center of the traveled spot. A spot showing "tailing" should be measured from the densest point of the spot. The distance of the solvent is measured from the base line to the solvent front.

$$R_f = \frac{\text{distance traveled by compound}}{\text{distance traveled by developing solvent}}$$

The R_f value for a certain compound is related to the nature of stationary phases and developing solvents, the temperature, the size of the sample spots, the thickness of TLC plates as well as the visualization methods, etc[204]. In general, a successful separation is achieved if all the R_f values of the components in the mixture are in the range of 0.30-0.60, and considerably different from one another[205].

- **Summarized Procedures for Thin-layer Chromatography Techniques**[206]

① Take a precoated TLC plate and gently draw the base line on it with a pencil.
② Spot the plate on the base line with a solution of the sample to be analyzed.
③ Develop the TLC plate in a developing chamber.
④ Visualize the spots on the TLC plate.
⑤ Calculate the R_f values for each point.

- **Questions**

① List three factors that will affect the R_f values of a compound in TLC analysis.
② Explain why the diameter of the spot on the TLC plate should be neither too small nor too large.
③ In a TLC separation, why must the level of the developing solvent be less than the height of the spots on the TLC plate?
④ In a TLC analysis, what visualization techniques can be used to visualize the spots of colorless compounds?
⑤ A mixture of the following three compounds (a)-(c) is separated by thin-layer chromatography, using silica gel as the adsorbent, and methylene dichloride as the developing sol-

---

[203] 在薄层色谱中,将溶质移动距离与展开剂移动距离的比值称为比移值(R_f)。
[204] R_f值通常与吸附剂的性质、色谱板的厚度、展开剂的性质、样品点的大小、温度及显色方法有关。
[205] 一般来说,如果混合物中各组分的R_f值都在0.30～0.60范围内,且彼此之间相差较大,则表明达到了比较理想的分离效果。
[206] 薄层色谱的操作步骤:①样品的溶解;②点样;③展开;④显色;⑤R_f值的计算。

vent. Predict the order of R_f values for compounds (a)-(c), and explain your answer.

⑥

        Ph-CH₂OH   Ph-CHO   Ph-COOH
          (a)   (b)   (c)

⑦ In a TLC analysis on a silica gel plate, if two compounds share the same R_f value ($R_f = 1$) when 2-propanol is used as the developing solvent, can it be concluded that they are identical compound? Explain your answer.

2. Column Chromatography

• **Principle**

Column chromatography is another kind of solid-liquid adsorption chromatography, the fundamental principle of which is the same as thin-layer chromatography. Although the column chromatography requires considerably more time than TLC in separation, it is efficient to separate and purify compounds in a larger scale from milligrams to hundreds of grams[207].

In column chromatography, the stationary phase is solid adsorbents (silica gel or aluminum oxide) packed into a column, while an eluting solvent serves as the mobile phase. When the eluting solvent is added from the top of the column, the sample initially adsorbed on the adsorbent at the top of column will begin to move down the column along with the eluting solvent. The interactions of the individual components of the sample with the stationary phase and the mobile phase determine the rate at which the different components elute from the column. In general, the more weakly the component is absorbed, the more rapidly it will be eluted. Thus, with a polar adsorbent, the less polar component will travel faster down the column than the more polar components[208]. The progressive separation of the mixture by column chromatography is hereby shown in Figure 2.20.

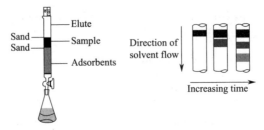

Figure 2.20 Separation of mixtures by column chromatography

---

    [207] 柱色谱也是一种固-液吸收色谱,其原理和薄层色谱相似。尽管柱色谱要比薄层色谱复杂且耗时,但柱色谱可用于较大量样品的分离。

    [208] 样品中的各组分与固定相和流动相之间的相互作用决定了不同组分从柱子上被洗脱下来的速度。一般来说,吸附能力较弱的组分,先随洗脱剂流出。因此,当使用极性吸附剂时,极性较弱的组分将比极性较强的组分优先被洗脱。

• Materials and Methods in Column Chromatography

(1) Adsorbents[209]

Silica gel and aluminum oxide are two common kinds of solid adsorbents used for column chromatography. Most column chromatography separations use silica gel as the adsorbents because of their moderate polarity and relatively high stability. The silica gel used in column chromatography separations must have a uniform particle size and high specific area. For a simple column chromatography, silica gel of a 70-230 mesh particle size is usually used. The activity of adsorbents depends on the amount of water containing. Column chromatographyic silica gel generally has 10%-20% absorbed water by weight.

(2) Eluting solvent

In column chromatography, the eluting solvent is used to dislodge the components on the top of the adsorbents in the column. The rates at which each component exits from the column are associated with the polarity of the eluting solvent. The proper choice of eluting solvents is the key to a successful column chromatography separation.

The selection of the eluting solvent is also a trial-and-error process for column chromatography, and the same criterion as the selection of a solvent in thin-layer chromatography could also be applied to column chromatography. Components in a sample moving down the column too rapidly with the eluting solvent will not be separated well from one another. In this case, another eluting solvent with less polarity should be tried. On the contrary, if the components still remain permanently adsorbed to the silica gel near the top of the column, while the eluting solvent moving down the column, the eluting solvent should be changed to a more polar one[210]. In general, a solvent moving the desired component in an approximate R_f value of 0.3 should be considered a good eluting solvent for column chromatography separation.

Column chromatography has high consistency with thin-layer chromatography when the same solid adsorbents are used as the stationary phase. Compared with column chromatography, the trials with thin-layer chromatography can be performed more quickly, and the amount of sample in need is smaller. To this end, thin-layer chromatography is always used as a forerunner to determine the eluting solvents for column chromatography separation[211]. The eluting solvents must be as pure as possible. Small quantities of impurities can dramatically alter the polarity of an eluting solvent. For instance, the presence of water in an organic solvent can significantly increase the polarity of the solvent[212].

---

[209] 吸附剂。

[210] 如果样品中的各组分洗脱的速度过快，而达不到很好的分离效果时，应重新选择另一种极性较小的溶剂作为洗脱剂。相反，如果当洗脱溶剂向下移动时，各组分仍吸附在柱顶附近的硅胶上很难被洗脱下来，则应更换极性较强的洗脱剂。

[211] 如果使用相同的固体吸附剂作为固定相，柱色谱与薄层色谱的分离效果基本一致。因而，可以用薄层色谱作为先导，通过实验来确定柱色谱要用的洗脱剂。

[212] 洗脱剂必须纯净。少量杂质的存在将会改变洗脱溶剂的极性。例如，洗脱剂中如果含有水，其极性将显著增加。

(3) Selection of a column

Once the appropriate adsorbent is selected, the amount of adsorbents to be used for the column chromatography separation should be decided, which, indeed, depends on the amount of the sample to be separated. In practical, a ratio of 20∶1 to 30∶1 between the amount of the adsorbents and that of the sample to be separated will be adopted for a moderate operation. In the case of a relatively similar polarity of each component in the mixture, the separation will be difficult and more adsorbents will be required[43].

A longer and thinner column will retain components longer on the stationary phase and a more polar eluting solvent is required to pull the components off the adsorbent, which may lead to the poor separation of each component of the sample due to the molecular diffusion. Usually, the height of 10-20 cm of silica gel is adopted and the ratio in the range of 8∶1～10∶1 between the height of the silica gel and the inside diameter of the column is normally adopted[44]. For a smaller R_f difference between the components in a sample, a column with a larger size is often required.

(4) Packing a column

The packing of a column is crucial to a successful column chromatography separation. A well-packed column should have a flat surface on the top of the adsorbent and be firmly packed with no air bubbles, cracks, and channels in the adsorbent[45]. The chromatographic column is usually a long glass tube with a stopcock at the bottom. The adsorbent can be added into the column by using a dry-packing or wet-packing method[46]. The general procedures for dry-packing method are as follows:

① Fix the dry and clean column onto an iron stand in a vertical position.

② Push a small piece of glass wool to the bottom of the column, and then cover the glass wool with about 5 mm of sand. Herein, the glass wool and sand are used as support materials to prevent the loss of silica gel through the stopcock.

③ Close the stopcock of the column and add the eluting solvent(or the less polar solvent)to half-height of the column. Then, open the stopcock and allow the solvent dropping into an Erlenmeyer flask.

④ Pour the requisite amount of silica gel slowly into the column through a funnel and keep the silica gel fall uniformly to the bottom.

⑤ Tamp the side of the column gently as the silica gel falling through the solvent to keep the top of the silica gel always be horizontal and to prevent the formation of air bubbles or channels in the adsorbent, which may lead to a poor separation. More fresh solvents may

---

⓸ 吸附剂的用量一般是待分离样品量的 20～30 倍。如果样品中各组分的极性比较接近，分离难度将增大，吸附剂的用量可适当增加。

⓹ 一般情况下，柱中吸附剂的高度为 10～20 cm，柱子的内径是吸附剂高度的 1/10～1/8。

⓺ 色谱柱内吸附剂的表面要平整，柱内吸附剂中没有明显的裂痕和气泡。

⓻ 装柱的方法有湿法和干法两种。

be needed to keep the solvent level always above the level of the silica gel at any time㊱.

⑥ Add 5 mm of quartz sand on top carefully after all the adsorbents are settled.

⑦ Apply pressure to pack the column firmly until only a small amount of solvent is above the sand.

⑧ Close the stopcock and get prepared for loading the sample to the column on the top of the adsorbent.

Columns can also be packed with the wet-packing method, and the general procedures for the wet-packing method are listed as:

① Put the requisite amount of silica gel slowly into an Erlenmeyer flask with an excess of solvent(about 1.5 times of the volume of silica gel). Petroleum ether and dichloromethane are often used to prepare the silica gel slurry. Then, mix the solvent and silica gel thoroughly and rigorously to remove the air present in the silica gel slurry㊳.

② Add the eluting solvent(or the less polar solvent)to half-height of the column.

③ Add the pre-mixed slurry of silica gel into the column in batches through a funnel. The slurry should be swirled thoroughly before adding each portion to the column.

④ Rinse and flush the slurry left in the Erlenmeyer flask into the column using a small amount of fresh solvent. The silica gel sticking to the wall of the column should be rinsed down as well.

⑤ Tap the side of the column constantly as the slurry flows down the column. Then, add 5 mm of quartz sand to the top of the silica gel carefully after all the silica gel is settled.

⑥ Apply pressure to pack the column firmly until only a small amount of solvent is left above the quartz sand. More fresh solvents may be needed to keep the solvent level always falling above the level of the adsorbent at any time.

⑦ Close the stopcock and get prepared for loading the sample to the column.

(5) Applying samples to the column㊴

A liquid sample can be applied onto the surface of silica gel in the column directly. A solid sample must be dissolved in a solvent before being applied onto the column, and the general procedures for applying a solid sample to the column is as follows:

① Dissolve the sample with a minimal volume of suitable solvents in appropriate polarity and make the solution as concentrated as possible. The total volume of the solution should not exceed 4-5 mL㊵.

② Apply the solution carefully to the column.

③ Rinse the container with a small amount of solvent and apply the left sample to the column.

㊱ 轻轻敲击柱身,使吸附剂慢慢而均匀地下沉,以保持吸附剂表面平整,且柱身无气泡或裂痕。操作过程中要始终保持柱内溶剂液面高于吸附剂的上层水平面。

㊳ 称取所需量的吸附剂,边搅拌边慢慢加入装有适量溶剂(石油醚或二氯甲烷)的锥形瓶中。加完后,充分搅拌混合物,以除去吸附剂中的空气。

㊴ 加样。

㊵ 将样品溶解在合适的溶剂中,溶剂的用量要尽量少,一般不超过 5 mL。

④ Add 1-2 mm of quartz sand to the top of the sample and prevent the sample and silica gel from being disturbed.

⑤ Open the stopcock and allow the upper level of the solvent just to be above the top of quartz sand.

⑥ Close the stopcock.

(6) **Elution of the sample**[21]

Elution of the components in the sample is performed by using a series of more progressively polar elution solvents. The less polar solvent is firstly used to elute the less polar components, and a more polar solvent or solvent combination is then adopted to elute the polar components. The polarity of the eluting solvent should be increased systematically and gradually. An abrupt change of the eluting solvent will cause cracks and channels in the adsorbent column. As the elution progress goes on, the components in the sample will be separated into a series of bands in the column[22]. The general procedures for a sample elution are illustrated as follows:

① Add the eluting solvent to the column slowly and carefully, so that the upper layer of the column will not be disturbed.

② Open the stopcock of the column and collect the eluent with a test tube or Erlenmeyer flask. Change the collection container in time when the current container is full.

③ Continuously add the eluting solvent to make sure that the column never dries out and maintains an optimum elution rate of 1-2 mL per minute. A too slow flowing rate will cause the diffusion of the component bands. In this case, compressed air can be applied at the top of the column to accelerate the flow rate of the eluting solvent[23].

④ Progressively increase the polarity of the eluting solvent to elute the components with high polarities quickly.

(7) **Collection of fractions**

If the components in the sample are colored, combine the eluent sharing the same color into a round-bottom flask. However, column chromatography separation is not limited to colored compounds. If the components are colorless, thin-layer chromatography analysis can be used to ascertain the purity of the eluent in each container. Combine the eluents that are identified as the same pure component into a round-bottom flask[24].

(8) **Recovery of pure components**

The pure separated components in the sample can be recovered by evaporating the elu-

---

[21] 洗脱。

[22] 洗脱过程中，洗脱剂的极性要逐渐增加。先用极性较弱的溶剂洗脱极性较弱的组分，然后换极性强的溶剂来洗脱极性较强的组分。随着洗脱过程的进行，样品中被分离的各组分将在色谱柱中呈现出一系列的色带。

[23] 不断添加洗脱剂，确保洗脱过程中柱子要始终被溶剂浸泡。洗脱液流出的速度以每分钟1～2 mL为宜。如果流速过慢，会造成样品色带的重叠。

[24] 如果样品中的组分是无色的，可以先用薄层色谱来确定每个接受容器中的流出液的纯度。合并含有相同组分的流出液并收集在同一个圆底烧瓶中。

ting solvents from the eluents collected in the round-bottom flasks. A rotary vacuum evaporator is often used for the removal of the solvents.

(9) Cleaning of columns

When all the components to be separated are respectively obtained from the column, drain out the remaining solvent in the column, then invert the column and push out the dried silica gel to a beaker and discard the silica gel to a solid-waste container.

- **Summarized Procedures for Column Chromatography Techniques**[25]

① Pack the column with the adsorbent.
② Apply the sample mixture to the top of the column.
③ Gradiently elute the absorbed components with eluting solvents.
④ Collect the different fractions from the bottom of the column.
⑤ Evaporate the eluting solvent to recover the pure separated components.

- **Questions**

① How to prevent air bubbles from being trapped in the adsorbent when you pack the column using the wet-packing method?

② A mixture of the three compounds(a)-(c) is to be separated by column chromatography using neutral aluminum oxide as the adsorbent and petroleum ether as the eluting solvent. Predict the order in which compounds(a)-(c) will elute from the column?

③ When a mixture of *syn*-azobenzene and *anti*-azobenzene is separated by column chromatography using neutral aluminium oxide as the adsorbent and methanol as the eluting solvent. Which isomer of zaobenzene will move down the column first? Explain your answer.

④ In the elution process of a sample, the absorbed components should be eluted with progressively more polar eluting solvents. Why must the eluting solvents be changed in an order from less polar solvent to a more polar one?

⑤ Why is it important to keep the level of the elution solvent not dropping below the top of the adsorbent in a column chromatography separation?

⑥ How will the following operations affect the results of column chromatography separation?

    a. A too strong adsorbent is adopted.
    b. A large amount of elution fractions are collected in one container.
    c. The flowing rate of the eluting solvent is very slow or very fast.

---

[25] 柱色谱的操作步骤：①装柱；②上样；③洗脱；④收集；⑤回收。

3. Gas Chromatography

Chromatography was discovered by Mikhail Tsvet (Russian botanist) in the early 1900s, which was used as a method for separation of an organic mixture. Compared with liquid-solid column chromatography, the gas chromatography (GC) offers an efficient chromatographic technique for separation and characterization of volatiles and semivolatiles in various analytical samples. Due to the advantages of low cost, low detection limit, fast analysis speed, and high separation efficiency, gas chromatography has been applied in various areas including petrochemicals, biochemicals, pharmaceuticals, and environmental protection, and even been used for tracing of persistent organic pollutants, forensic science, fire residues, and detection and quantitation of alcohol in drunk-driving cases. In addition, the gas chromatography can be used by combining with mass spectrometry as established analytic technique with many applications in the detection and measurement of low concentration analytes.

Gas chromatography contains one stationary phase and one mobile phase. The mobile phase is always an inert gas with low or negligible adsorption capacity, such as nitrogen, helium, or hydrogen, while the stationary phase is either liquid or solid. When a solid is applied as the stationary phase, it is called gas-solid chromatography (GSC). In comparison, the liquid as a stationary phase is called gas-liquid chromatography (GLC). For example, the gas-liquid chromatography may employ the inert nitrogen as the mobile gas phase and high boiling point liquid adsorbed onto a solid as the stationary phase. By carrying the analyte with the mobile phase through the heated chromatographic column, the individual components could be gradually separated.

When a mixture of substances is injected at the column inlet, the components carried through the column would be dissolved, adsorbed, or coordinated on the stationary phase surface, and then re-carried by the carrier-gas to realize the separation of components from each other. Thus the separation depends on sample components partition between the stationary phase and the mobile phase according to the distribution constant (K) on the basis of the strength of the intermolecular forces between the solute and the surface of the stationary phase. Therefore, a component with a large distribution constant will stay more strongly in the stationary phase than that with a smaller distribution constant. In this case, the former will take longer time to pass through the chromatographic column to reach the detector than the latter. This time starting from the column inlet to the detector is called the retention time (t_R) and is characteristic of the chemical properties of the solute and the stationary phase. Like all chemical equilibria, K is highly dependent on temperature, so retention time is temperature dependent.

However, there are some limitations for the type of compounds amenable for GC analysis. They have to be easily vaporized without decomposing or reacting with the components of stationary and mobile phases or with other components present in the sample to be analyzed. To be suitable for analysis by GC, compounds should have a relatively higher vapor pressure, generally below 350-400℃. Most of the compounds can be directly analyzed by their current form, however, some others need to be transformed into their derivatives featuring high volatile, less polar, and/or more thermally stable.

As shown in Figure 2.21, gas chromatographic equipment commonly consists of a carrier gas system, injector, vaporizing chamber, gas chromatographic column, temperature control system, detector and data processing unit. When the analytes are injected through the injector port, the components are immediately vaporized and carried by the mobile phase to flow through the chromatographic column. The components are then separated into stationary phases according to the different strength of the intermolecular forces between the solute and the surface of the stationary phase. The separated components can then be detected and displayed on data acquisition unit.

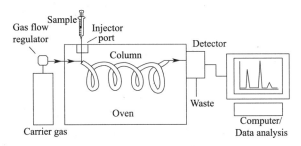

Figure 2.21   Gas chromatographic equipment

- **Questions**

① What is retention time?

② What's the advantages and disadvantages of gas chromatography?

③ What's the similarities and differences between gas chromatography and column chromatography?

④ Can gas chromatography be used for analyzing the sample containing some components at high-boiling point?

4. High Performance Liquid Chromatography

High performance liquid chromatography(HPLC) was introduced in the 1970s to distinguish this high-pressure technique from the traditional low-pressure column chromatography. As a modern application of liquid chromatography(LC), HPLC is emerged as a powerful tool for sensitive, specific, and precise analysis, which can solve several shortcomings of liquid chromatography, such as slow separation and low efficiency. Since a smaller and narrower column has been utilized for HPLC, high pressure was required to force the mobile phase carrying with the sample through this chromatographic column and a more effective separation could be achieved in much less time. Despite these advantages, special HPLC equipment must be of high quality to make sure of the reproducible results, rendering expensive for this technique.

Compared with gas chromatography limited to volatile samples, soluble compounds even the involatile, thermally stable, and large molecules are all well-suitable for HPLC. In this chromatography, samples can be analyzed without any pre-treatment such as chemical derivatization and can be recovered after analysis due to a non-destructive technique applied. Nowadays, HPLC has become one of the most popular and powerful separation techniques to separate, identify, and quantify the active compounds in a mixture, which makes this technique extensively applied in the areas of chemicals, pharmaceuticals, agriculture, and so on. In addition, the mobile phase is always liquid, commonly n-hexane and isopropanol for normal phase HPLC and acetonitrile and methanol for reverse phase HPLC, while the stationary phase is either liquid or solid. When a solid is applied as the stationary phase, it is called liquid-solid chromatography(LSC), in comparison, the liquid as a stationary phase is called liquid-liquid chromatography(LLC). Furthermore, the combination of high performance liquid chromatography with mass spectrometry provides an important analytic technique for separation and characteristic at the same time.

The chromatographic column used in HPLC is narrow and packed with tiny particles(3 to 5 μm in diameter). When a small volume of liquid sample is injected at the top of chromatographic column, the liquid mobile phase carries the analytes through the column packed with stationary phase forced by high pressure derived from a pump toward the detector. During the process, distribution between mobile phase and stationary phase through surface absorption, ion exchange, relative solubilities and steric affect results in effective separation of the components in the mixture. Then the detector will display the retention times of each component in the sample(shown as peaks by data acquisition system). Retention time mainly depends on the interactions between stationary phase, mobile phase, column temperature, and the components themselves. The amount of each component can be calculated by internal-standard or external-standard methods. For chiral HPLC analysis, the ratio of peak areas of enantiomeric isomers demonstrates the enantiomeric ratio of this compound.

HPLC can be divided into normal phase or reverse phase types. For normal phase HPLC, a polar stationary phase and a non-polar mobile phase such as n-hexane and isopropanol are utilized, thus the analytes can be separated based on polarity. Therefore, this method can be applied to water-sensitive compounds and chiral compounds. In contrast, reverse phase HPLC employs nonpolar material as the stationary phase and moderate polar solvent as the mobile phase, which is used to separate non-polar, polar, ionizable, and ionic molecules. Before the sample being analyzed, calibration standards are necessarily interspersed by blank injections of mobile phase to ensure cleanliness of the instrument and reach a flat baseline.

As shown in Figure 2.22, instrument for HPLC commonly contains solvent reservoir(mobile phase), pump(packed with stationary phase), sample injector, columns, column oven, detector, and data acquisition system. The solvent for HPLC should be pure enough because of narrow columns packed with tiny stationary phase being applied. Chromatographically pure solvents are needed and should be degassed before use. A pump providing about 400 atm pressure is necessary to force mobile phase carrying the solute through HPLC column successfully. Automated or single(manual) sample injectors are needed to make sure of adding sample solution with a high level of accuracy. HPLC columns should be compatible with high pressure, which are made of stainless steel, with 50 and 300 mm long and internal diameter between 2 and 5 mm. Column oven is used to maintain a constant column temperature, which then keeps a good reproducibility for analysis. Detectors could be UV-spectroscopy, fluorescence, mass-spectrometric or electrochemical detectors. Signals from the detector may be collected and displayed on data acquisition system. The retention time and peak area for each component can be recorded in the computer.

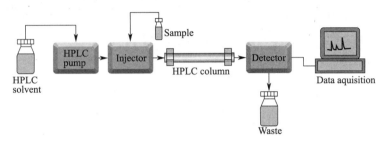

Figure 2.22 Instrument for high performance liquid chromatography

- **Questions**

① What are the differences between liquid chromatography, high performance liquid chromatography, and gas chromatography?

② What are the types of HPLC? Point out the differences.

③ Why is high pressure needed for HPLC?

④ What solvent is used in HPLC?

⑤ What are the advantages of HPLC compared with GC?

Chapter 3
Fundamental Experiments

3.1 Unsaturated Hydrocarbons

Exp.1 Preparation of Cyclohexene

- **Objectives**

1. To study the principle and method for the preparation of alkenes by acid-catalyzed dehydration of alcohols.
2. To learn to perform a miniscale reaction.
3. To practice the technique of fractional distillation.
4. To practice the operation of washing and drying of liquid compounds.
5. To learn the method for the purification of liquid compounds by distillation.

- **Principle**

Alkenes can be prepared by intramolecular dehydration of alcohols, and the reaction is usually performed in the presence of an acid catalyst, such as concentrated sulfuric acid or phosphoric acid. During the reaction, the acid can provide protons to activate alcohols and effectively promote the removal of water from the intermediate protonated alcohol. For example, cyclohexene can be prepared by the dehydration of cyclohexanol using 85% phosphoric acid as the catalyst. Phosphoric acid, a moderate acid, is hereby employed to prevent extensive charring[20].

$$\text{C}_6\text{H}_{11}\text{OH} \underset{\triangle}{\overset{H_3PO_4}{\rightleftharpoons}} \text{C}_6\text{H}_{10} + H_2O$$

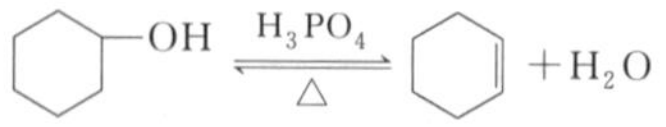

The acid-catalyzed dehydration reaction of cyclohexanol is reversible. In order to shift the equilibrium to the right and maximize the yield of cyclohexene, a fractional distillation is applied to distill cyclohexene out of the reaction system upon its formation[21].

The intermolecular dehydration of cyclohexanol is one of the commonly competitive re-

---

[20] 环己烯通常用环己醇在磷酸催化下分子内脱水的方法来制备。用比较缓和的磷酸代替硫酸作为催化剂，可以减少反应过程中炭化的发生。

[21] 此反应为可逆反应。为了使平衡向产物方向移动，提高环己烯的收率，通常采用分馏的方法将生成的环己烯不断地从反应体系中蒸馏出来。

actions during the reaction process, when, the reaction temperature should be carefully controlled to minimize this side reaction[20].

$$\text{C}_6\text{H}_{11}\text{OH} \xrightleftharpoons[\Delta]{\text{H}_3\text{PO}_4} \text{C}_6\text{H}_{11}\text{-O-C}_6\text{H}_{11} + \text{H}_2\text{O}$$

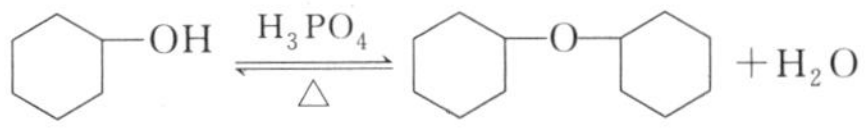

- **Apparatus**

Apparatuses required include 15 mL round-bottom flask, fractionating column, Liebig condenser, distillation head, thermometer, thermometer adapter, receiver adapter, Erlenmeyer flask, 5 mL graduate cylinder, beaker and separatory funnel.

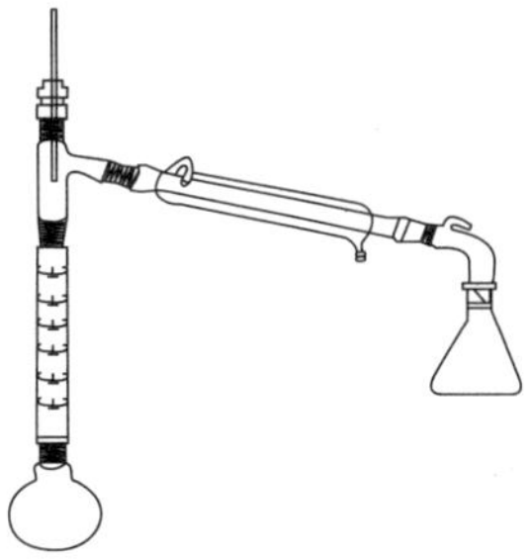

- **Setting Up**

Place 5.0 mL of cyclohexanol and several zeolites in the 15 mL round-bottom flask. Add 2.5 mL of 85% phosphoric acid, and then mix the liquids thoroughly by gently swirling the flask[21]. Install the apparatus for fractional distillation as illustrated in Figure 3.1.

Figure 3.1 Apparatus for fractional distillation

- **Experimental Procedures**

1. Heat the reaction mixture in the flask to reflux gently and collect the distillates while maintaining the temperature below 85 ℃[22].

2. Stop heating when there are only about 2.5 mL of stick liquids in the reaction flask[23].

3. Transfer the distillate into a separatory funnel. Stand the funnel until the content in the funnel is separated into two distinct layers, and collect the organic layer.

4. Wash the organic layer with 5.0 mL of saturated aqueous brine solution[24] and remove the aqueous layer.

5. Transfer the organic liquid into a clean and dry Erlenmeyer flask. Dry the solution over several amount of anhydrous calcium chloride in portions. Occasionally swirl the Erlenmeyer flask until the solution becomes clear. Then, after 10 min, add additional portions of anhydrous calcium chloride if calcium chloride clumps together or the liquid remains cloudy.

6. Decant the dried organic solution into a dry 15 mL round-bottom flask and assemble the flask for simple distillation.

7. Collect the fraction boiling between 82-85 ℃ (760 mmHg) in a pre-weighed receiving flask.

8. Weigh the product and calculate the percentage yield of cyclohexene.

---

[20] 控制反应的温度，可以减少副反应的发生。
[21] 反应物要充分混合均匀后再加热，以避免炭化的发生。
[22] 环己烯的沸点为 83 ℃，环己烯与水共沸，沸点为 70.8 ℃，环己醇与水形成共沸物的沸点为 97.8 ℃。所以分馏时需控制好温度，使柱顶温度不要超过 85℃，以确保在环己烯蒸出的同时，又能减少未作用的环己醇的蒸出。
[23] 反应终点的确定可参考以下现象：①反应进行了 40 min 左右；②反应瓶内出现白雾；③柱顶温度下降后又升到 85 ℃ 以上。
[24] 用饱和食盐水洗涤有机相的目的是减少产物在水相中的溶解，同时使分层现象更明显。

• **Helpful Hints**

1. Phosphoric acid is corrosive. Wear gloves while handling this reagent.

2. Cyclohexene is toxic and irritating. Wear gloves and avoid skin contact while handling it. Perform this reaction in the hood to minimize exposure to it.

3. Cyclohexene and water can form azeotrope with a boiling point of 70.8 ℃, while cyclohexanol and water can also form azeotrope with a boiling point of 97.5 ℃. Perform the fractional distillation carefully to keep the temperature below 85 ℃.

4. Washing the organic layer with a solution of saturated sodium chloride not only reduces the solubility of the organic product in the aqueous layer, but also simplifies the delamination.

5. Make sure that all the glassware used for the final distillation is clean and dry[23].

6. In the final distillation step, decant the dried organic solution to the distilling flask carefully to prevent the drying agent from falling into the flask[24].

7. The boiling point of cyclohexene is 82-85 ℃ (760 mmHg).

• **Questions**

1. What measures should be taken to improve the yield of cyclohexene in this reaction?

2. Why is the temperature of the top of the fractionating column kept below 85 ℃ in the initial step of the dehydration reaction?

3. How should the ending point of this reaction be determined? Whether the produced water reaching to the theoretical amount can be used as the indicator for the end of the reaction?

4. Why is it particularly important that the crude cyclohexene should be dried enough prior to the final distillation?

Exp.2 Preparation of Phenylacetylene

• **Objectives**

1. To study the principle and method of converting an alkene into an alkyne by the combination of halogenation and dehydrohalogenation.

2. To practice the assembling of the reflux apparatus with mechanical stirring.

3. To practice extraction of liquid compounds.

4. To practice the rotary evaporation for the removal of volatile solvents.

5. To learn the principle and operation of vacuum distillation.

• **Principle**

Alkynes can be prepared by the elimination of hydrogen halide from alkyl halides in a similar manner as alkenes. Treatment of a 1,2-dihaloalkane with an excess amount of a strong base can result in the sequential elimination of two molecules of hydrogen halide and

---

[23] 最后一步蒸馏所用的玻璃仪器使用前必须放入烘箱干燥，以免蒸馏时出现前馏分，影响产品的纯度。

[24] 向蒸馏烧瓶内转移干燥后的液体时，要小心不要把干燥剂倒入烧瓶内。

lead to the formation of an alkyne. 1,2-Dihaloalkane can be easily obtained by the addition reaction of halogen with an alkene. Thus, halogenation of an alkene followed by the double dehydrohalogenation provides us a useful method to convert an alkene into the corresponding alkyne[25].

$$RCH=CH \xrightarrow{X_2} R-\underset{H}{\overset{X}{C}}-\underset{H}{\overset{X}{C}}-R \xrightarrow{B^-} \underset{R}{\overset{X}{C}}=\underset{H}{\overset{R}{C}} \xrightarrow[\Delta]{B^-} R-C\equiv C-R$$

The mechanism of this conversion involves three steps, electrophilic addition of halogen across an alkene to form 1,2-dihaloalkane, base-induced elimination of hydrogen halide to give an intermediate haloalkene, and the second elimination of hydrogen halide from vinyl dihalide. Among which, the last step is somewhat difficult to occur, therefore a strong base and more forcing conditions are always required.

In this experiment, phenylacetylene, a reaction synthon in organic synthesis, is prepared by the reaction of phenylethylene with bromine and then subsequent base-induced elimination of two molecules of hydrogen bromide[26]. This conversion process is shown as follows:

$$\text{PhCH=CH}_2 \xrightarrow[\text{CCl}_4]{\text{Br}_2} \text{PhCHBrCH}_2\text{Br} \xrightarrow[\text{methanol}]{\text{KOH}} \text{PhC}\equiv\text{CH}$$

• **Apparatus**

Apparatuses required include 100 mL three-necked flask, reflux condenser, addition funnel, Büchner funnel, suction flask, separatory funnel, 100 mL round-bottom flask, distillation head, Liebig condenser, thermometer, thermometer adapter, receiver adapter, Erlenmeyer flask, graduate cylinder, rotary evaporator and apparatus for mechanical stirring and vacuum distillation.

• **Setting Up**

Place 8 g of phenylethylene and 50 mL of carbon tetrachloride into a 100 mL three-necked round-bottom flask, and assemble the reaction apparatus shown in Figure 3.2.

• **Experimental Procedures**

1. Preparation of 1,2-dibromophenylethane

(1) Add 20 mL of carbon tetrachloride and 15 g of bromine into the addition funnel.

(2) Cool the mixture in the flask in an ice-water bath, and then add the solution of bromine in carbon tetrachloride dropwise into the flask with vigorous stirring.

---

[25] 烯烃的卤化和随后的两次卤化氢消除的组合为我们提供了一个将烯烃转化为相应炔烃的方法。

[26] 苯乙炔是重要的有机合成中间体。本实验采用苯乙烯与溴加成后，在强碱的作用下消除两当量的溴化氢的合成路线来制备苯乙炔。

(3) Continue to stir the reaction mixture for one hour after the addition[27].

(4) Remove carbon tetrachloride with the rotary evaporator, collect the remaining white crystals in the flask, and then have them air-dried.

2. Preparation of phenylacetylene

(1) Add 15 g of newly obtained 1,2-dibromophenylethane and 12 g of potassium hydroxide into a 100 mL three-necked round-bottom flask.

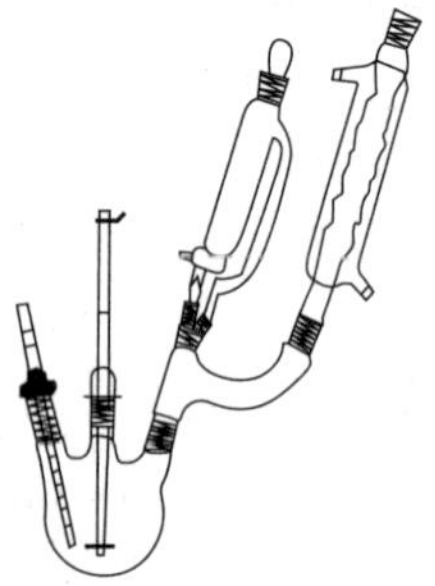

Figure 3.2   Apparatus for the preparation of phenylacetylene

(2) Place 15 mL of methanol into the addition funnel, and then add methanol to the reaction mixture in the flask slowly.

(3) Heat the mixture to reflux with stirring for 1.5 h upon the completion of the addition.

(4) Stop heating and stirring, then cool the reaction mixture to room temperature and remove the solids by suction filtration.

(5) Transfer the filtrate into a separatory funnel and extract it with diethyl ether twice.

(6) Collect the ether extracts in a round-bottom flask, and then assemble the flask for simple distillation to remove methanol.

(7) Assemble the apparatus for vacuum distillation, and distill and collect the pure phenylacetylene product at 42-43 ℃/18 mmHg.

(8) Weigh the final product and calculate the percentage yield of phenylacetylene.

- **Helpful Hints**

1. Bromine is a hazardous chemical. Perform the reaction of bromine in the hood to minimize exposure.

2. Potassium hydroxide is corrosive. Wear gloves while handling it and avoid skin contacting with potassium hydroxide. Wash the affected area with copious amounts of cool water if your skin contacts with potassium hydroxide[28].

3. Methanol is flammable and volatile. Be sure that there are no open flames anywhere in the laboratory. Use methanol in a well-ventilated area and avoid prolonged contact or inhalation[29].

4. Be careful to wear proper eye protection during the vacuum distillation.

5. The elimination of hydrogen bromide from 1,2-dibromophenylethane is violently exothermic, so the methanol must be added slowly to avoid vigorous reaction, which may result

---

[27] 苯乙烯与溴加成反应的最佳反应温度为 12 ℃ 左右。若温度过低，反应瓶内的液体会十分黏稠，难以搅拌。

[28] 氢氧化钾具有腐蚀性，操作时请戴手套，避免皮肤直接接触氢氧化钾。如果不小心皮肤接触到了氢氧化钾，要马上用大量的冷水清洗患处。

[29] 甲醇易燃、易挥发，应在通风良好的地方使用，且确保实验室不能有明火。要尽量避免长时间接触或吸入甲醇。

in the evaporation of methanol[20].

6. The reaction time should be controlled between 1.5-2 h, since prolonged reaction time will reduce the yield of phenylacetylene due to the increased side reaction of triple bond at high temperature[21].

7. The optimal reaction temperature for the first bromine addition reaction is about 12 ℃. If the temperature is too low, the reaction solution in the flask will be too stick to be stirred smoothly.

8. Although the boiling point of phenylacetylene is 142-143 ℃, some side reactions of triple bond will occur when phenylacetylene is heated for a long time. Therefore, phenylacetylene should be generally purified by vacuum distillation rather than simple distillation[22].

• **Questions**

1. What measures should be taken to improve the purity of phenylacetylene in this experiment?

2. The boiling point of phenylacetylene is 142-143 ℃. Why is it best to purify phenylacetylene by vacuum distillation rather than simple distillation?

3. Discuss the differences observed in the IR and NMR spectra of phenylacetylene and phenylethylene.

4. Consult the relevant literature and discuss the advantages and disadvantages of the reported routes for the synthesis of phenylacetylene.

3.2 Alkyl Halides

Exp.3 Preparation of 1-Bromobutane

• **Objectives**

1. To study the method of preparing primary alkyl halides from primary alcohol using hydrohalic acid.

2. To learn the assembling of the reflux apparatus with a gas-trap.

3. To practice extraction, washing and distillation of organic liquid compounds.

• **Principle**

The common preparing method for alkyl halides is to treat the corresponding alcohols with hydrohalic acid. This reaction with primary alcohols works best via the $S_N 2$ pathway, and heating must be used to ensure the completion of the reaction. For instance, *n*-butyl bromide can be prepared in the laboratory by treating *n*-butyl alcohol with hydrobromic acid. Hydrobromic acid is corrosive and suffocated, and may cause hazards to health when people

---

[20] 消除卤化氢的一步是剧烈放热的，因此甲醇应缓慢滴加，避免反应过于剧烈而发生喷料。

[21] 消除反应这步的反应时间应控制在1.5～2h，反应时间过长苯乙炔的收率反而会降低，这可能是因为三键长时间加热会发生其他的副反应。

[22] 苯乙炔的沸点是142～143 ℃，但分子中的三键长时间加热会发生副反应，所以一般用减压蒸馏来提纯苯乙炔。

are exposed to it. On the other hand, hydrobromic acid can be easily obtained by mixing sodium bromide with concentrated sulfuric acid. Accordingly, the preparation of *n*-butyl bromide in this experiment is simply achieved by treating *n*-butyl alcohol with sulfuric acid and sodium bromide[23]. The main reactions are shown as below:

$$NaBr + H_2SO_4 \longrightarrow NaHSO_4 + HBr$$

$$HBr + n\text{-}C_4H_9OH \xrightarrow{\triangle} n\text{-}C_4H_9Br + H_2O$$

The reaction of *n*-butyl alcohol with hydrobromic acid is reversible, so a large excess of hydrobromic acid is hereby used to drive the conversion to the right.[24]

Accompanied by main reactions, the following side-reactions may readily occur:

$$2HBr + H_2SO_4 \xrightarrow{\triangle} Br_2 \uparrow + SO_2 \uparrow + 2H_2O$$

$$CH_3CH_2CH_2CH_2OH \xrightarrow[\triangle]{H_2SO_4} CH_3CH_2CH=CH_2 + CH_3CH=CHCH_3$$

$$2CH_3CH_2CH_2CH_2OH \xrightarrow[\triangle]{H_2SO_4} (CH_3CH_2CH_2CH_2)_2O + H_2O$$

To minimize these side reactions and improve the yield of *n*-butyl bromide, the following measures are taken: (1) dilute sulfuric acid is employed to avoid the oxidation of hydrobromic acid, as well as the formation of by-products; (2) gentle heating is adopted to control the reaction temperature and prevent the formation of by-products 1-butene (or 2-butene) and butyl ether during the intramolecular dehydration of *n*-butanol. Moreover, reflux apparatus with a gas-tap system is applied for the absorption of poisonous gases generated during reactions.[25]

- **Apparatus**

Apparatuses required include 100 mL round-bottom flask, 50 mL round-bottom flask, Allihn condenser, long-stem funnel, thermometer adapter, graduate cylinder, separatory funnel, thermometer, distillation head, Liebig condenser, adapter, Erlenmeyer flask, beaker and stopper.

- **Setting Up**

Place 8.3 g of sodium bromide in the 100 mL round-bottom flask, and add 6.2 mL of *n*-butyl alcohol and several zeolites. Mix the contents of the flask thoroughly by swirling the flask. Then, add cold dilute sulfuric acid (10 mL concentrated sulfuric acid + 10 mL water) to the flask with swirling slowly. Install the reflux apparatus with a gas-trap as illustrated in Figure 3.3.

- **Experimental Procedures**

1. Warm the flask gently until most of the sodium bromide have dissolved, and then heat

---

[23] 氢溴酸具有腐蚀性和窒息性，长时间接触对身体有害。在本实验中，通过溴化钠与浓硫酸作用得到氢溴酸，氢溴酸一经形成，就接着与正丁醇发生亲核取代反应来生成1-溴丁烷。

[24] 正丁醇与氢溴酸的反应是可逆的，为了使平衡右移，实验中氢溴酸是过量的。

[25] 为了减少副反应的发生，提高1-溴丁烷的收率，在实验中一般采取以下措施：（1）用63%的硫酸代替浓硫酸，防止氢溴酸的氧化和副产物的生成；（2）控制加热的速度和反应温度，防止反应物正丁醇脱水生成正丁醚、1-丁烯。

the mixture under gentle reflux for about 30 min while shaking the flask frequently[26].

2. Stop heating, and allow the mixture in the flask to be cooled down.

3. Add several zeolites to the mixture again when it is no longer boiling, and then assemble the flask for simple distillation(Figure 3.4)[27].

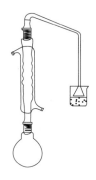

Figure 3.3 The reflux apparatus with a gas-trap

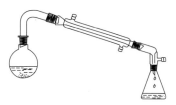

Figure 3.4 Apparatus for simple distillation

4. Distill the mixture rapidly until the distillate appears to be clear[28].

5. Transfer the distillate to a separatory funnel.

6. Shake the separatory funnel well, let it stand for a moment until the content in the funnel is separated into two distinct layers, and then collect the organic layer.

7. Transfer the saved organic layer to another dry separatory funnel, wash the organic layer with 3 mL of concentrated sulfuric acid, and remove the acid layer.

8. Wash the organic layer sequentially[29] with 10 mL of water, 5 mL of 10% sodium carbonate solution and 10 mL of water.

9. Transfer the obtained organic layer to a dry and clean Erlenmeyer flask, and dry it with a small amount of anhydrous calcium chloride[30].

10. Swirl the flask occasionally for 10-15 min until the solution is clear. Add further small portions of anhydrous calcium chloride if the drying agent clumps together or if the solution still appears cloudy.

11. Transfer the well-dried solution to a clean and dry 50 mL round-bottom flask carefully, and then assemble the flask for distillation after several zeolites are added to the flask.

12. Collect the fraction boiling between 99-102 ℃(760 mmHg)in a pre-weighed dry receiving flask.

13. Weigh the final product and calculate the percentage yield of *n*-butyl bromide.

---

[26] 反应过程中，要保持溶液缓慢沸腾，且经常摇动反应瓶，以免发生炭化和副反应，影响1-溴丁烷的收率。

[27] 这一步蒸馏的目的是将粗产物1-溴丁烷蒸出，便于后处理。同时，随着1-溴丁烷的蒸出，反应也更趋于完全。

[28] 1-溴正丁烷要完全蒸出，否则会影响产率。可以从以下三个方面来判断蒸馏是否结束：①馏出液是否由浑浊变为澄清；②蒸馏烧瓶内上层油层是否消失；③用盛有少量水的烧杯收集馏出液，观察有无油滴出现。

[29] 依次洗涤。

[30] 用无水氯化钙来干燥粗产物，不仅可以除去少量的水，还可以除去粗产物中痕量的正丁醇，起到进一步提纯的作用。否则正丁醇会和1-溴丁烷形成共沸物（b.p.＝98.6 ℃）而难以除去，影响产物的纯度。

• **Helpful Hints**

1. Concentrated sulfuric acid is corrosive and can cause severe burns. Be extremely careful while handling it. If any concentrated sulfuric acid comes in contact with your skin, immediately wash it off with copious amounts of cold water and then with dilute sodium bicarbonate solution[①].

2. Alkyl halides are toxic, and some alkyl bromides are lachrymators and suspected carcinogens. Perform this reaction in the hood, and wear gloves to avoid skin contact with n-butyl bromide.

3. The reactants must be mixed thoroughly by shaking the flask to reduce extensive charring and other side reactions, which will result in lower yield of n-butyl bromide.

4. Detect the leak of the separation funnel carefully and make sure that the separatory funnel is well enough before washing organic compounds with it[②].

5. Concentrated sulfuric acid is used to remove unreacted n-butyl alcohol as well as the by-products 1-butene(or 2-butene) and dibutyl ether from crude n-butyl bromide. This step must be performed in a dry separatory funnel[③].

6. Make sure that you are working with the correct layer that contains the desired product while performing the washing operation[④]. The following methods can help you to determine which layer is the organic layer and which is the aqueous layer:

(1) Compare the density of the extracting solvent and water. Usually, the one with a larger density will be on the bottom.

(2) Add 1-2 mL of water to the separatory funnel, and the layer on which the water drop stays must be the aqueous layer.

(3) Take out 1-2 drops of liquid from one of the layers and put it onto a watch glass, and then add several drops of water. The layer must be the organic layer if the liquid on the watch glass appears cloudy.

7. It is recommendable to keep all the layers obtained from different washing steps. Never discard any layer until the experiment is completed[⑤].

• **Questions**

1. The concentration of sulfuric acid matters considerably for the conversion of n-butyl alcohol to n-butyl bromide. What will be the adverse effects if concentrated sulfuric acid is used for this reaction?

2. When crude n-butyl bromide is transferred to a separatory funnel after the first distillation, the organic layer is normally in the bottom layer, but it sometimes remains in the top

---

① 稀的碳酸氢钠溶液。
② 分液漏斗使用前一定要认真检漏,以免在洗涤过程中造成溶液的渗漏。
③ 浓硫酸是用来洗去粗产物中未反应的正丁醇和副产物1-丁烯、2-丁烯和正丁醚。这一步的操作要在干燥的分液漏斗中进行,以确保硫酸的浓度。
④ 每一次的分液操作前,都要仔细判断你需要的有机相是在上层还是下层。可通过比较有机物和水的比重,或向分液漏斗缓慢滴加1~2 mL水来区分水相和有机相。
⑤ 为保证洗涤过程中,产物不会被错误地丢弃,建议将每次分液后的液体都保留,直到洗涤过程全部结束。

layer. Try to analyze the reason.

3. What is the purpose of each washing operation in the work-up procedure of this experiment?

(1) Washing with concentrated H_2SO_4.

(2) Washing with water.

(3) Washing with 10% Na_2CO_3.

(4) Washing with water again.

4. 1-Butene (or 2-butene) and dibutyl ether are both by-products of this reaction and can be removed by concentrated sulfuric acid. Write down the corresponding reaction equation.

5. Reddish-brown color is sometimes observed in the organic layer after being washed with concentrated sulfuric acid. Why? What should be done next?

6. Why must crude n-butyl bromide be dried completely prior to the final distillation? Can solid sodium hydroxide or potassium hydroxide be used to replace calcium chloride and dry crude n-butyl bromide? Explain the reason.

Exp.4   Preparation of 2-Chloro-2-methylbutane

- **Objectives**

1. To study the preparing principle of tertiary alkyl halides from tertiary alcohols using concentrated hydrohalic acids.

2. To learn to perform a reaction using a separatory funnel.

3. To practice washing the liquid compounds.

4. To practice the drying and simple distillation of liquid compounds.

- **Principle**

The most efficient method for the preparation of tertiary alkyl halides is to treat the tertiary alcohols with concentrated hydrohalic acids. Tertiary alcohols can react with either HCl or HBr at 0 ℃ by a S_N1 mechanism. Take the preparation of 2-chloro-2-methylbutane for instance, it is synthesized in the laboratory by treating 2-methyl-2-butanol with hydrochloric acid.

$$CH_3CH_2-\underset{\underset{CH_3}{|}}{\overset{\overset{CH_3}{|}}{C}}-OH + HCl \longrightarrow CH_3CH_2-\underset{\underset{CH_3}{|}}{\overset{\overset{CH_3}{|}}{C}}-Cl + H_2O$$

This reaction takes place when acid protonates the hydroxyl oxygen atom to yield oxonium ions. Water is then expelled to generate a carbocation intermediate, and the carbocation will then react with nucleophilic chloride ion to give the product㉖.

In the sequence, the formation of the carbocation is the slow and rate-determining step (rds). The formed intermediate carbocation can not only react with chloride ion in the S_N1 pathway but also can cause the loss of a proton from the neighboring carbon atom to give a

---

㉖ 在酸的作用下，羟基质子化形成䦛盐。接着脱去水，形成中间体碳正离子。最后，碳正离子与亲核性的氯负离子反应得到最终产物。

$$CH_3CH_2-\underset{\underset{CH_3}{|}}{\overset{\overset{CH_3}{|}}{C}}-OH \xrightarrow{HCl} CH_3CH_2-\underset{\underset{CH_3}{|}}{\overset{\overset{CH_3}{|}}{C}}-\overset{+}{O}H_2 \xrightarrow{-H_2O} CH_3CH_2-\underset{\underset{CH_3}{|}}{\overset{\overset{CH_3}{|}}{C}}+ \xrightarrow{Cl^-} CH_3CH_2-\underset{\underset{CH_3}{|}}{\overset{\overset{CH_3}{|}}{C}}-Cl$$

mixture of alkenes in the E1 pathway[27]. The elimination reaction of 2-methyl-2-butanol to produce corresponding alkenes is always competed with the reaction in which nucleophilic substitution is performed to form 2-chloro-2-methylbutane.

$$CH_3CH_2-\underset{\underset{CH_3}{|}}{\overset{\overset{CH_3}{|}}{C}}-OH \xrightarrow{HCl} CH_3CH_2-\underset{CH_3}{\overset{CH_2}{C}} + CH_3CH=\underset{CH_3}{\overset{CH_3}{C}}$$

- **Apparatus**

Apparatuses required include 50 mL round-bottom flask, 125 mL separatory funnel, an ice-water bath, long-stem funnel, thermometer, thermometer adapter, graduate cylinder, distillation head, Liebig condenser, adapter, Erlenmeyer flask and stopper.

- **Setting Up**

Place 10 mL of 2-methyl-2-butanol and 25 mL of concentrated(12 mol/L) hydrochloric acid in a separatory funnel(Figure 3.5).

Figure 3.5 The separatory funnel

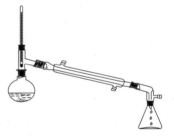

Figure 3.6 Apparatus for simple distillation

- **Experimental Procedures**

1. Mix the reactants by swirling the separatory funnel gently for about 1 min without the stopper on the funnel.

2. Place the stopper securely, and then invert the funnel. Open the stopcock of the funnel immediately to release the excess pressure in the funnel.

3. Close the stopcock and shake the funnel for 5 min while frequently venting[28].

4. Stand the funnel until the content in the funnel is separated into two distinct layers, and then remove the acidic layer.

---

[27] 在整个反应进程中，生成碳正离子的一步是反应的决速步骤。生成的碳正离子，不仅可以与氯负离子发生亲核取代反应生成卤代烃，同时也会发生 E1 消除生成相应的烯烃。

[28] 为了防止分液漏斗内压力过大，分液漏斗振摇后一定要放气。

5. Wash the organic layer sequentially with 10 mL of saturated sodium chloride solution, 10 mL of cold saturated aqueous sodium bicarbonate[28] and finally 10 mL of cold water.

6. Remove the aqueous layers carefully, and transfer the crude 2-chloro-2-methylbutane to a dry and clean Erlenmeyer flask.

7. Dry the crude product with a small amount of anhydrous calcium chloride. Swirl the Erlenmeyer flask occasionally and let it stand for 10-20 min until the solution is clear. Then, add further small portions of anhydrous calcium chloride if the drying agent clumps together or if the solution still appears cloudy.

8. Transfer the dried crude solution to a clean and dry 50 mL round-bottom flask carefully. Add several zeolites to the flask, and then assemble the flask for distillation (Figure 3.6) with the receiver flask immersing an ice-water bath[29].

9. Collect the fraction boiling between 83-86 ℃ (760 mmHg) in the receiving flask[30].

10. Weigh the obtained product and calculate the percentage yield of 2-chloro-2-methylbutane.

- **Helpful Hints**

1. Detect the leak of the separatory funnel carefully and make sure that the separatory funnel is well enough before performing this reaction.

2. Concentrated hydrochloric acid fume is suffocating, so handle it in the hood. If any acid spills on your skin, wash it off with large amount of water and then with dilute sodium bicarbonate solution.

3. Alkyl halides are toxic. Work in the hood and wear gloves to avoid skin contact. Do not breathe alkyl halide vapors.

4. Vent the separatory funnel to avoid a buildup of pressure during the initial step of the reaction.

5. Saturated aqueous solution of sodium bicarbonate is used to remove the acid still present in the organic layer. Considering the generation of CO_2 gas with high pressure, venting is also necessary in this step[31].

6. Keep the drying agent anhydrous calcium chloride from falling into the distillation flask while transferring dried 2-chloro-2-methylbutane to the flask for distillation[32].

7. Collect the pure 2-chloro-2-methylbutane maintaining the temperature at 83-86 ℃ (760 mmHg). The fraction distilled out above 90 ℃ may be a mixture of the product and impurities.

---

① 饱和碳酸氢钠溶液。

② 2-氯-2-甲基丁烷的沸点为 83~86 ℃，蒸馏时最好将接收瓶置于冰水浴中。

③ 最后一步蒸馏提纯时，要严格控制收集馏分的沸点范围（83~86 ℃）。90 ℃以上蒸馏出的馏分可能是产品和杂质的混合物。

④ 碳酸氢钠的饱和溶液用来洗去有机相中残留的酸。在这一步操作中，由于酸碱中和反应会生成二氧化碳气体，使分液漏斗中压力增大，一定要及时排气。

⑤ 蒸馏前要把干燥剂过滤干净。许多无机盐的水合物在温度超过 30~40 ℃时，会失水而分解。

• **Questions**

1. What is the purpose of each washing operation in the work-up procedure of this experiment?

(1) Washing with saturated sodium chloride solution.

(2) Washing with cold saturated aqueous sodium bicarbonate.

(3) Washing with water.

2. The crude 2-chloro-2-methylbutane is dried with anhydrous calcium chloride in this reaction. Can anhydrous calcium chloride be replaced by solid sodium hydroxide or potassium hydroxide to dry crude products? Explain the reason.

3. Consider the by-products formed by E1 processes in the reaction of 2-methyl-2-butanol with HCl, which of the following two alkenes would be preferred? Explain the reason.

$$CH_3CH_2-\underset{\underset{CH_3}{|}}{\overset{\overset{CH_3}{|}}{C}}-OH \xrightarrow{HCl(aq)} CH_3CH_2-\underset{\underset{CH_3}{|}}{C}=CH_2 + CH_3CH=\underset{\underset{CH_3}{|}}{\overset{\overset{CH_3}{|}}{C}}$$

4. Discuss the differences observed in the IR and ^{1}H NMR spectra of 2-methyl-2-butanol and 2-chloro-2-methylbutane that are consistent with the conversion of an alcohol into an alkyl chloride in this procedure.

3.3 Alcohols, Phenols and Ethers

Exp.5 Preparation of 2-Methyl-2-butanol

• **Objectives**

1. To study the principle and method of preparing alcohols via Grignard reactions.

2. To learn to perform a water-sensitive and air-sensitive reaction.

3. To practice the extraction, washing and drying of liquid compounds.

4. To learn to distill low-boiling point liquids.

5. To learn the method for the purification of liquid compounds by distillation.

• **Principle**

Organomagnesium halides (RMgX), also known as Grignard reagents, can be prepared by the reaction of alkyl halides with elemental magnesium.

$$R(Ar)-X + Mg \xrightarrow{dry\ ether} R(Ar)-MgX \quad X=Cl, Br, I$$

The carbon-magnesium covalent bonds in Grignard reagents are polar, making the carbon atoms bear partial negative charge basic and highly nucleophilic[20]. Therefore, aprotic and nonelectrophilic solvents such as diethyl ether or tetrahydrofuran (THF) must be used for the preparation of Grignard reagents. Besides, ethereal solvents can form complexes with Gri-

---

[20] 格氏试剂中 C—M 键的极化，使得和镁相连的碳原子带有部分的负电荷，具有一定的碱性和亲核性。

gnard reagents, thus stabilizing the Grignard reagents[25]. Grignard reagents are water-sensitive and air-sensitive, its reactions must therefore be performed under an inert atmosphere, and all the reagents, solvents, and apparatus to be used for Grignard reactions must be dry[26].

Grignard reagents can serve as nucleophiles in many organic reactions. For example, the nucleophilic addition reactions of Grignard reagents with aldehydes or ketones are commonly used method to synthesize various kinds of secondary or tertiary alcohols[27]. In this experiment, ethylmagnesium bromide is firstly prepared via the reaction of ethyl bromide with magnesium metal in anhydrous diethyl ether, and then the formed ethylmagnesium bromide nucleophilically attacks acetone to produce the magnesium alkoxide intermediate, followed by an acid workup for the final product of 2-methyl-2-butanol.

$$CH_3CH_2-Br + Mg \xrightarrow{\text{dry ether}} CH_3CH_2-MgBr$$

$$CH_3CH_2-MgBr + O=C\begin{matrix}CH_3\\CH_3\end{matrix} \xrightarrow{\text{dry ether}} H_3C-\underset{OMgBr}{\overset{CH_3}{\underset{|}{\overset{|}{C}}}}-CH_2CH_3$$

$$H_3C-\underset{OMgBr}{\overset{CH_3}{\underset{|}{\overset{|}{C}}}}-CH_2CH_3 \xrightarrow{H_3^+O} H_3C-\underset{OH}{\overset{CH_3}{\underset{|}{\overset{|}{C}}}}-CH_2CH_3 + Mg(OH)Br$$

Figure 3.7 Apparatus for the preparation of 2-methyl-2-butanol

• **Apparatus**

Apparatuses required include 150 mL three-necked flask, Allihn condenser, addition funnel, separatory funnel, drying-tube containing anhydrous calcium chloride, 50 mL round-bottom flask, thermometer, thermometer adapter, distillation head, Liebig condenser, adapter, Erlenmeyer flask, graduate cylinder, beaker and apparatus for magnetic stirring.

• **Setting Up**

Weigh 1.8 g of freshly crushed magnesium turnings and place them in the three-necked flask containing a stir bar. Assemble the apparatus as shown in Figure 3.7. Then, add 10 mL of anhydrous diethyl ether to the reaction flask.

• **Experimental Procedures**

1. Preparation of ethylmagnesium bromide

(1) Place 10 mL of dry ethyl bromide and 8.0 mL of anhydrous diethyl ether in the addi-

---

[25] 制备格氏试剂所需的溶剂通常是醚类溶剂，如：乙醚或四氢呋喃。这些溶剂还可以通过与格氏试剂形成配合物，对格氏试剂起到一定的稳定作用。

[26] 格氏试剂对水和空气敏感，所以格氏试剂的合成和相关反应应在无水、无氧的条件下进行，所用仪器和试剂要绝对干燥。

[27] 格氏试剂具有亲核性，它与醛酮的亲核加成反应是实验室合成仲醇和叔醇十分常用的方法。

tion funnel, and swirl the solution gently.

(2) Add dropwise approximately 5.0 mL of ethereal solution of ethyl bromide from the addition funnel to the flask. The solution will soon begin to boil, with small bubbles forming on the surface of the magnesium turnings, and the reaction mixture will become slightly cloudy[㉘].

(3) Heat the mixture gently for several minutes or add a small crystal of iodine to the flask to initiate the reaction if the reaction does not start spontaneously after 10 min[㉙].

(4) Turn on the stirrer and add the rest of ether solution of the ethyl bromide dropwise from the addition funnel to the reaction mixture at a rate that maintains a gentle reflux.

(5) When all the ether solution of the ethyl bromide has been added, continue to stir the mixture under gentle reflux for 10-15 min until the disappearance of most of the magnesium metal.

(6) Cool the resulting ether solution of ethylmagnesium bromide with an ice-water bath for direct use in the following step without further treatment.

2. Preparation of 2-methyl-2-butanol

(1) Place the mixture of 5.0 mL of anhydrous acetone and 5.0 mL of anhydrous diethyl ether in the addition funnel and swirl the solution gently.

(2) Add the ether solution of acetone dropwise to the stirred solution of ethylmagnesium bromide.

(3) Allow the resulting mixture to be stirred for 15 min at room temperature after all the ether solution of acetone is added.

(4) Mix 3.0 mL of concentrated sulfuric acid and 10 mL of water in a beaker, and transfer the dilute solution of sulfuric acid into the addition funnel[㉚].

(5) Cool down the reaction flask with an ice-water bath and add the dilute sulfuric acid dropwise to the reaction mixture in the flask while stirring.

(6) Transfer the resulting solution to a separatory funnel upon the completion of the addition, separate the layers, and collect the organic layer in an Erlenmeyer flask.

(7) Extract the aqueous layer with 10 mL of diethyl ether twice, and combine these ether extracts with the saved organic layer.

(8) Transfer the combined organic layer to a separatory funnel and wash it with 8 mL of 10% sodium carbonate aqueous solution.

(9) Remove the aqueous layer and transfer the organic layer to a dry and clean Erlen-

---

㉘ 随着反应液加热回流，镁带表面有气泡生成，反应液由澄清变为灰色浑浊，表明反应已经开始。

㉙ 如果十分钟后还没有明显的反应现象，可以用水浴温热反应液或向反应液中加入少量的碘来促使反应引发。

㉚ 随着硫酸溶液的滴加，反应液逐渐澄清。如果硫酸滴加结束后，溶液中还有少量白色固体，可补加适量稀硫酸溶液直至溶液完全澄清。

meyer flask. Then, dry the ether solution with anhydrous potassium carbonate. Swirl the flask occasionally for 10-15 min until the solution in the Erlenmeyer flask becomes clear.

(10) Carefully decant the dried ether solution into a dry round-bottom flask, and then assemble the flask for simple distillation to remove diethyl ether.

(11) Continue to heat the solution and collect the fraction boiling between 100-104 ℃ (760 mmHg) in a pre-weighed receiving flask.

(12) Weigh the final product and calculate the percentage yield of 2-methyl-2-butanol.

● **Helpful Hints**

1. All the glassware used for the preparation of Grignard reagent and the following nucleophilic addition reaction using Grignard reagents should be dried in the oven at 110℃ at least for one hour before use.

2. Diethyl ether and ethyl bromide are extremely flammable and volatile, so make sure that there are no open flames anywhere in the laboratory. Use liquids at a low boiling point in a well-ventilated area and avoid prolonged contact or inhalation.

3. Concentrated sulfuric acid is corrosive and can cause severe burns. Be extremely careful while handling it. If any concentrated sulfuric acid comes in contact with your skin, immediately wash it off with copious amounts of cold water and then with dilute sodium bicarbonate solution.

4. Alkyl halides are toxic, and some alkyl bromides are lachrymators and suspected carcinogens. Perform the reaction in the hood. Wear gloves and avoid skin contact.

5. Check all ground-glass joints in the apparatus carefully and mate them tightly to prevent the escape of diethyl ether and other substances with a low boiling point during the reaction[①].

6. The preparation of the ethylmagnesium bromide, the nucleophilic addition of ethylmagnesium bromide with acetone, and the following quenching of the reaction by dilute acid are all exothermic. Therefore, reaction reagents should be slowly added to avoid the danger to be possibly caused by too vigorous reaction[②].

7. A small quantity of iodine can be added to facilitate the reaction of ethyl bromide and magnesium turnings.

8. The obtained solution of ethylmagnesium bromide should be used at once in the subsequent nucleophilic addition reaction to avoid the following side reactions[③].

$$RMgX + H_2O \longrightarrow RH + Mg(OH)X$$

[①] 在反应过程中,要确保所有玻璃仪器间的接口都结合紧密,以防低沸点的乙醚和溴乙烷的逸出。

[②] 格氏试剂的合成反应以及后续的亲核加成反应和最后的酸化步骤,都是放热的,因而反应试剂的滴加速度应加以控制。如果反应过于剧烈,要用冰水浴冷却反应液,以免造成试剂的逸出或副反应的发生。

[③] 格氏试剂可以与空气中的氧气、二氧化碳和水发生反应。因而,格氏试剂一经合成,应立即进行下一步的反应,不可久置。

$$RMgX + 1/2 O_2 \longrightarrow ROMgX \xrightarrow{H_2O} ROH + Mg(OH)X$$

$$RMgX + CO_2 \longrightarrow RCO_2MgX \xrightarrow{H_2O} RCOOH + Mg(OH)X$$

9. 2-Methyl-2-butanol can form an azeotrope with a boiling point of 87℃. The crude 2-methyl-2-butanol should be dried thoroughly prior to the final distillation to prevent the fraction from being distilled out below 95 ℃, which will affect the purity and percentage yield of the 2-methyl-2-butanol㉗.

- **Questions**

1. Grignard reagents are air-sensitive and water-sensitive. What measures should be taken to create water-free and air-free experimental conditions for the preparation of ethylmagnesium bromide and the following nucleophilic addition in this experiment?

2. In the preparation of ethylmagnesium bromide, why should the rest ether solution of ethyl bromide be added dropwise to the reaction mixture from the addition funnel only when the Grignard reaction has been initiated?

3. Why should anhydrous rather than solvent-grade diethyl ether be used to prepare the solution of acetone used for the nucleophilic addition reaction.

4. Besides the desired ethylmagnesium bromide, a small quantity of by-product butane is also obtained from the reaction of ethyl bromide with magnesium turnings. Explain the reason. What measurement should be taken to minimize this side reaction?

5. Why can anhydrous calcium chloride not be used for drying crude 2-methyl-2-butanol prior to the final distillation?

Exp.6　Preparation of Triphenylmethanol

- **Objectives**

1. To study the principle and method of preparing tertiary alcohols bearing two identical substituents by the reaction of Grignard reagents with esters.

2. To learn to perform a water-sensitive and air-sensitive reaction.

3. To practice the extraction, washing and drying of organic liquid compounds.

4. To learn the method of dealing with low-boiling point liquids.

5. To practice the recrystallization of solid compounds.

- **Principle**

Grignard reagents, an important class of organometallic compounds, are commonly prepared by the reactions of alkyl or aryl halides with magnesium metal in the ether solvent. For instance, phenylmagnesium bromide can be derived from bromobenzene in diethyl ether according to the following equation:

㉗　2-甲基-2-丁醇会与水形成共沸物，为了确保产物的纯度，提高其收率，粗产物在蒸馏提纯前必须严格干燥。

$$\text{C}_6\text{H}_5\text{Br} + \text{Mg} \xrightarrow{\text{diethyl ether}} \text{C}_6\text{H}_5\text{MgBr}$$

The carbon-magnesium covalent bonds in Grignard reagents are polar. The carbon atom connecting with the magnesium is nucleophilic, and can thus react readily with many electrophiles including alkyl halides, carbonyl compounds and epoxides, etc. The nucleophilic substitution reaction of esters with 2 equivalents of a Grignard reagent is a good method to prepare a tertiary alcohol in which two of the substituents are identical[25]. In this experiment, triphenylmethanol is synthesized by the reaction of phenylmagnesium bromide with methyl benzoate as follows:

$$2\ \text{C}_6\text{H}_5\text{MgBr} + \text{C}_6\text{H}_5\text{C(O)OMe} \xrightarrow[\text{H}_3\text{O}^+]{\text{diethyl ether}} \text{C}_6\text{H}_5-\underset{\underset{\text{C}_6\text{H}_5}{|}}{\overset{\overset{\text{OH}}{|}}{\text{C}}}-\text{C}_6\text{H}_5$$

The reaction occurs following the usual nucleophilic substitution mechanism. The first equivalent of phenylmagnesium bromide is added to methyl benzoate, and the loss of methoxy magnesium bromide from the tetrahedral intermediate yields a benzophenone. Then a second equivalent of phenylmagnesium bromide is immediately added to benzophenone, followed by an acid workup to produce triphenylmethanol[26].

Grignard reagents are water-sensitive and air-sensitive, and the formation of phenylmagnesium bromide and the subsequent nucleophilic substitution reaction involving phenylmagnesium bromide must therefore be performed under an inert atmosphere. Meanwhile, all the reagents, solvents, and apparatus to be used for the reactions involving phenylmagnesium bromide must be dry.

• **Apparatus**

Apparatuses required include 100 mL three-necked flask, Allihn condenser, addition funnel, drying-tube containing anhydrous calcium chloride, thermometer, thermometer adapter, distillation head, Liebig condenser, adapter, Erlenmeyer flask, Büchner funnel, filter flask, graduate cylinder, beaker and apparatus for magnetic stirrer.

---

[25] 两摩尔格氏试剂和酯的亲核取代反应是有机化学实验室合成带有两个相同取代基的叔醇常用方法。

[26] 反应大致分为三步：①苯基格氏试剂与苯甲酸酯发生亲核取代，失去甲氧基溴化镁，得到二苯甲酮；②二苯甲酮与第二摩尔的苯基格氏试剂发生加成反应，生成正四面体的加成产物；③加成中间体在酸性条件下水解，得到三苯甲醇。

• **Setting Up**

Weigh 0.5 g of freshly crushed magnesium turnings and place them in a three-necked flask that contains a stir bar. Assemble the apparatus for magnetic stirring as shown in Figure 3.8.

• **Experimental Procedures**

1. Preparation of phenylmagnesium bromide

Figure 3.8　Apparatus for the preparation of triphenylmethanol

(1) Place 2.1 mL of bromobenzene and 10 mL of anhydrous diethyl ether in the addition funnel, and swirl the solution gently.

(2) Add dropwise approximately 2 mL of the ethereal solution of bromobenzene from the addition funnel to the flask. The solution will soon begin to boil with small bubbles forming on the surface of the magnesium turnings, and the reaction mixture will become slightly cloudy㉗.

(3) If the reaction does not start spontaneously after 5 min, warm the reaction mixture gently for several minutes or add a small crystal of iodine to the flask to initiate the reaction.

(4) Turn on the stirrer and add the rest of ether solution of the bromobenzene dropwise from the addition funnel to the reaction mixture at a rate that maintains a gentle reflux㉘.

(5) When all the ether solution of the bromobenzene is added, heat the mixture in a warm-water bath and keep the mixture refluxing gently for 30 min until most of the magnesium metal has disappeared.

(6) Cool the resulting ether solution of phenylmagnesium bromide with an ice-water bath for direct use in the following step.

2. Preparation of triphenylmethanol

(1) Place the mixture of 5.0 mL of anhydrous diethyl ether and 1.4 mL of ethyl benzoate in the addition funnel, and swirl the solution gently.

(2) Add the ether solution of ethyl benzoate dropwise from the addition funnel to the ether solution of phenylmagnesium bromide while stirring.

(3) Heat the resulting mixture in a warm-water bath at 50 ℃ and reflux the mixture gently for 30 min upon the completion of the addition.

(4) Place the reaction flask in a cold-water bath, and then add 15 mL of saturated solution of ammonium chloride dropwise from the addition funnel to the flask while stirring㉙.

(5) Upon the completion of the addition, assemble the reaction flask with a simple distillation apparatus to distill out diethyl ether until the appearance of the crystals of triphenylmethanol.

㉗　随着反应液加热回流，镁带表面有气泡生成，反应液由澄清变为浑浊，表明反应已经开始。

㉘　反应引发后，才能从滴液漏斗中开始滴加剩余的溴苯的乙醚溶液。要控制滴加的速度，避免反应过于剧烈，造成溶剂的损失和副产物联苯的生成。

㉙　加入饱和氯化铵溶液的目的是使二苯甲酮与苯基格氏试剂的加成物水解，得到最终产物三苯甲醇。

(6) Add 15 mL of petroleum ether to the reaction flask and stir the mixture for several minutes[20].

(7) Cool down the mixture with an ice-water bath to complete the crystallization process of triphenylmethanol.

(8) Collect the crude triphenylmethanol by vacuum filtration.

(9) Triphenylmethanol can be purified by recrystallization using a mixed solvent of petroleum ether and ethanol with the volume ratio of 2 : 1.

- **Helpful Hints**

1. All the glassware to be used during the formation and reaction of phenylmagnesium bromide should be dried in the oven at 110 ℃ at least for one hour before use.

2. Diethyl ether is extremely flammable and volatile, so be sure that there are no open flames anywhere in the laboratory. Use diethyl ether in a well-ventilated area and avoid prolonged contact or inhalation.

3. Bromobenzene and ethyl benzoate are irritants, so avoid contact with these liquids and do not breathe their vapors.

4. Check all ground-glass joints in the apparatus carefully and mate them tightly to prevent the escape of diethyl ether during the reaction.

5. A small crystal of iodine can be added to promote the initiation of the reaction between bromobenzene and magnesium turnings.

6. The ether solution of bromobenzene should be added slowly to keep the reaction mixture refluxing gently with the vapor-liquid conversion level at the half of the first ball of the Allihn condenser[21].

7. More biphenyl by-product will be formed in the case of too vigorous reaction of bromobenzene with magnesium turnings. Most of the biphenyl can be removed by being dissolved in the petroleum.

8. The obtained solution of phenylmagnesium bromide should be used at once in the subsequent nucleophilic substitution reaction to avoid the following side reactions:

$$RMgX + H_2O \longrightarrow RH + Mg(OH)X$$

$$RMgX + \frac{1}{2}O_2 \longrightarrow ROMgX \xrightarrow{H_2O} ROH + Mg(OH)X$$

$$RMgX + CO_2 \longrightarrow RCO_2MgX \xrightarrow{H_2O} RCOOH + Mg(OH)X$$

9. The role of ammonium chloride solution is to quench the addition reaction of benzophenone with phenylmagnesium bromide to yield the crude triphenylmethanol.

- **Questions**

1. Grignard reagents are air-sensitive and water-sensitive. What measures should be taken to create water-free and air-free experimental conditions for the formation and reaction

---

[20] 联苯易溶于石油醚中,通过此步骤可除去副产物联苯。

[21] 格氏试剂的合成过程中,要控制溴代苯的滴加速度,确保冷凝中气-液转换平面不超过冷凝管中第一个球的二分之一。

of phenylmagnesium bromide in this experiment?

2. In the preparation of phenylmagnesium bromide, why should the rest ether solution of bromobenzene be added to the reaction mixture only when the Grignard reaction is initiated?

3. The crude product contains the desired phenylmagnesium bromide and also small quantities of biphenyl by-product. What measures should be taken to minimize the formation of biphenyl? How should the generated biphenyl be removed in the subsequent work-up procedure?

4. Why is the solution of ammonium chloride instead of sulfuric acid used to decompose the addition product of benzophenone with phenylmagnesium bromide mixture?

5. Library project: In addition to the hereby used synthetic pathway, devise other synthetic pathways using Grignard reagents for the preparation of triphenylmethanol.

Exp.7 Preparation of Di-n-butyl Ether

- **Objectives**

1. To study the principle and method of synthesizing symmetrical ether via intermolecular dehydration of alcohol.

2. To learn the zaeotropic removal of water using water segregator.

3. To practice assembling the reflux apparatus with a water segregator.

4. To practice the washing, drying and distillation of liquid compounds.

- **Principle**

Simple symmetrical ethers can be prepared via the acid-catalyzed intermolecular dehydration of the corresponding primary alcohols[26]. The reaction occurs by the $S_N 2$ pathway usually involving three steps: (1) protonation of alcohols; (2) nucleophilic substitution of protonated alcohols; (3) acid-base reaction to form ethers[27]. The mechanism of this reaction is shown as below:

$$R-\overset{..}{O}H \xrightarrow{H^+} R-\overset{+}{O}H_2 \underset{S_N 2}{\overset{R-\overset{..}{O}H}{\rightleftharpoons}} R-\overset{+}{\underset{H}{O}}-R \overset{H_2\overset{..}{O}}{\rightleftharpoons} R-O-R$$

The intermolecular dehydration of alcohols is undertaken under acidic conditions. The acidic catalyst protonates the hydroxyl group of an alcohol to yield a good leaving group as water, which facilitates the following nucleophilic substitution of another equivalent of alcohol[28]. It should be noted that this method is limited to primary alcohols, while secondary and tertiary alcohols are easily dehydrated by an E1 mechanism to yield alkenes under the same

---

㉖ 伯醇在酸性条件下分子间脱水是实验室合成简单的对称单醚常用的方法。

㉗ 反应一般按照 $S_N 2$ 机理进行，大致分为以下三步：（1）醇羟基的质子化；（2）质子化的醇和另外一摩尔醇的亲核取代；（3）酸碱反应生成醚。

㉘ 醇分子间脱水生成醚的反应必须在酸性条件下进行。酸性催化剂提供质子，使醇羟基质子化，从而转化成一个好的离去基团水，以便后面的亲核取代反应。

conditions. For instance, dibutyl ether is obtained by the intermolecular dehydration of *n*-butyl alcohol using sulfuric acid as the catalyst.

$$2 \ \text{CH}_3\text{CH}_2\text{CH}_2\text{CH}_2\text{OH} \xrightleftharpoons[134\sim135\,^\circ\text{C}]{\text{H}_2\text{SO}_4} \ \text{CH}_3\text{CH}_2\text{CH}_2\text{CH}_2\text{O}\text{CH}_2\text{CH}_2\text{CH}_2\text{CH}_3 + \text{H}_2\text{O}$$

The intermolecular dehydration of *n*-butyl alcohol is reversible, and measures such as increasing the amount of one reactant or removing one of the products from the reaction mixture are therefore required to drive the equilibrium toward completion. Herein, azeotropic distillation is applied to remove the newly formed water as quickly as possible to shift the equilibrium to the right [25].

The intramolecular dehydration of alcohols to form alkenes is the main competing reaction along the intermolecular dehydration. The former usually takes place at a higher temperature. For instance, 1-butene will be the major product of the dehydration of *n*-butyl alcohol when the reaction temperature is higher than 135 °C [26]. Therefore, the reaction temperature should be carefully controlled to minimize side reactions.

$$\text{CH}_3\text{CH}_2\text{CH}_2\text{CH}_2\text{OH} \xrightleftharpoons[>135\,^\circ\text{C}]{\text{H}_2\text{SO}_4} \ \text{CH}_3\text{CH}_2\text{CH}=\text{CH}_2$$

- **Apparatus**

Apparatuses required include 100 mL three-necked flask, thermometer, thermometer adapter, water segregator, Allihn condenser, 50 mL round-bottom flask, distillation head, Liebig condenser, adapter, Erlenmeyer flask, stopper, separatory funnel, graduate cylinder and beaker.

- **Setting Up**

Place 31 mL of *n*-butyl alcohol and several zeolites in a 100 mL three-necked flask, and then add 5 mL of concentrated sulfuric acid into the reaction flask. Thoroughly mix the reaction mixture by gently swirling the flask, and assemble the reaction apparatus as shown in Figure 3.9.

- **Experimental Procedures**

1. Add water into the water segregator until the level is 2-3 mm lower than the top of its side neck.

2. Heat the reaction mixture slowly to reflux.

3. Continue to heat the mixture to 134-135 °C, and reflux it for approximately one hour until the water segregator is filled with water [27].

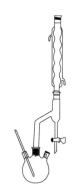

Figure 3.9 Apparatus for the preparation of dibutyl ether

㉕ 本实验中，采用共沸除水的方法，及时将生成的水脱离反应体系，使平衡右移，以提高正丁醚的产率。

㉖ 在酸性条件下，醇分子间脱水生成醚的反应和分子内脱水生成烯烃的反应是一对竞争反应。高温条件有利于烯烃的生成。

㉗ 反应过程中要严格控制反应的温度和时间。若加热时间过长，温度过高，会有炭化和副反应发生。

4. Stop heating and allow the reaction mixture to be cooled down to room temperature.

5. Transfer the mixture into a separatory funnel containing 50 mL of water. Shake the funnel thoroughly and let it stand for a moment until the content in the funnel is separated into two distinct layers.

6. Remove the water layer and wash the organic layer with 50％ sulfuric acid twice[28]. Then wash it with 10 mL of water twice.

7. Transfer the organic layer to a dry and clean Erlenmeyer flask, and dry it with anhydrous calcium chloride. Swirl the flask occasionally for 10-15 min until the solution in the flask becomes clear.

8. Transfer the dried solution to a clean and dry 50 mL round-bottom flask carefully. Add several zeolites and assemble the flask for distillation.

9. Collect the fraction boiling between 140-144 ℃(760 mmHg)in a pre-weighed receiving flask.

10. Weigh the obtained product and calculate the percentage yield of dibutyl ether.

• Helpful Hints

1. Concentrated sulfuric acid is corrosive and can cause severe burns. Be extremely careful while handling it. If any concentrated sulfuric acid comes in contact with your skin, immediately wash it off with copious amounts of cold water and then with dilute sodium bicarbonate solution.

2. The surface of the liquid in the side neck of water segregator will increase gradually with the evaporation and condensation of n-butyl alcohol, dibutyl ether and newly formed water. The densities of n-butyl alcohol and dibutyl ether are both less than water and form the upper layer of the liquid. Most of the n-butyl alcohol and dibutyl ether can overflow back into the reaction flask when the separator is full of liquid[29].

3. Observe the level of the liquid in the side neck of water segregator carefully. When the surface of the liquid is higher than the top of side neck, open the stopcock of water segregator and drip a few drops of water[30].

4. The reaction temperature and time must be controlled carefully to avoid extensive charring as well as the side reactions.

5. n-Butyl alcohol can dissolve in 50％ sulfuric acid, while dibutyl ether has little solubility in acid. In this case, unreacted n-butyl alcohol can be removed from the crude product by washing the organic layer with 50％ sulfuric acid.

6. The azeotropic boiling point of dibutyl ether and water is 92.9 ℃, so the crude prod-

---

[28] 正丁醇可以溶于50％的硫酸中,而正丁醚在酸中的溶解度很小。因此,借助这步操作可除去没有反应完的正丁醇。

[29] 反应生成的水、正丁醚和未反应的正丁醇受热共沸蒸出,遇冷凝管被冷凝后进入分水器,并在分水器中分层。其中,上层主要是正丁醇和正丁醚,下层主要是水。利用分水器可以使上层有机物不断地流回反应瓶,而将生成的水除去。

[30] 反应过程中,要认真观察分水器中的液面。当液面高于分水器支管位置时,打开分水器的开关放出几滴水,以确保分水器中的油层始终为薄薄的一层。当分水器中的水面和分水器支管相齐时,表明反应已经结束。

uct should be dried thoroughly prior to the final distillation[29].

• **Questions**

1. What measures should be taken to improve the yield of dibutyl ether in this experiment?

2. What measures should be taken to minimize the side reactions in this experiment?

3. What is the purpose of each washing operation in the work-up procedure of this experiment?

(1) Washing the reaction mixture with 50% H_2SO_4 twice.

(2) Washing the reaction mixture with water twice.

4. How much water will be collected from the side neck of water segregator after the reaction is completed. In general, the amount of water collected in the actual operation is always more than the theoretical value. Explain the reason.

5. Discuss the differences observed in the IR and ^1H NMR spectra of n-butyl alcohol and diethyl ether.

Exp.8 Preparation of Ethyl Phenyl Ether

• **Objectives**

1. To study the principle and method of preparing unsymmetrical ether via Williamson ether synthesis.

2. To practice assembling the reflux apparatus with mechanical stirrer.

3. To practice washing, extraction and drying of liquid compounds.

4. To learn the technique of vacuum distillation.

• **Principle**

The most generally useful method of preparing ethers involves Williamson ether synthesis, which uses an alkoxide ion or phenoxide as a nucleophile to attack alkyl halide or tosylate by an S_N2 pathway to form the ether bond[29]. This method is used not only to synthesize symmetric ethers but is also powerful to make mixed ethers that cannot be prepared via intermolecular dehydration of alcohols. The Williamson ether synthesis is carried out under basic conditions, and the main competitive reaction is the elimination reaction of alkyl halides to form corresponding alkenes. In general, the more hindered the alkyl halides, the more likely they are to undergo elimination reactions under basic conditions. Therefore, Williamson ether synthesis is generally applied with methyl halides or primary halides[29]. Herein, ethyl phenyl ether is synthesized by the reaction of phenol and ethyl bromide in the presence of sodium

---

[29] 正丁醚可以和水形成共沸物（b.p.=92.9 ℃）。因此，粗产物在最后蒸馏前应彻底干燥，以免前馏分影响产品正丁醚的纯度和收率。

[29] 威廉森合成法是合成醚（单醚和混合醚）最有效的方法。在此方法中，烷氧基负离子或酚氧基负离子作为亲核试剂进攻卤代烃，发生取代反应从而得到醚。

[29] 威廉森合成法通常以甲基的卤代烃或伯卤代烃为原料来合成醚。空间位阻大的卤代烃（仲卤代烃或叔卤代烃）在相同的条件下易发生消除反应生成烯烃。

hydroxide.

$$\text{C}_6\text{H}_5\text{OH} + \text{C}_2\text{H}_5\text{Br} \xrightarrow{\text{NaOH}} \text{C}_6\text{H}_5\text{OC}_2\text{H}_5 + \text{NaBr}$$

Sodium phenolate has low solubility in the organic phase because of its ionic nature. In order to ensure the complete interaction among the reactants and improve the efficiency of the heterogeneous reaction, vigorous stirring must be applied[24].

- **Apparatus**

Apparatuses required include 100 mL three-necked flask, Allihn condenser, addition funnel, 50 mL round-bottom flask, distillation head, Liebig condenser, thermometer, thermometer adapter, adapter, Erlenmeyer flask, graduate cylinder and apparatus for mechanical stirring and vacuum distillation.

- **Setting Up**

Add 7.5 g of phenol, 4.8 g of sodium hydroxide and 4 mL of water into a 100 mL three-necked round-bottom flask. Mix the reactants thoroughly, and then add several zeolites into the flask. Assemble the reaction apparatus as shown in Figure 3.10.

Figure 3.10　Apparatus for the preparation of ethyl phenyl ether

- **Experimental Procedures**

1. Add 8.5 mL of ethyl bromide into the addition funnel.

2. Heat the reaction mixture slowly while stirring until the solid is dissolved.

3. Add ethyl bromide dropwise to the reaction mixture while keeping the temperature at 80-90 ℃. The addition should be completed in about 1 h[25].

4. Continue to stir the resulting mixture at 80-90 ℃ for another two hours upon the completion of the addition.

5. Stop heating and stirring. Allow the reaction mixture to cool down to room temperature, and then add enough water to dissolve the solids in the flask.

6. Transfer the solution to the separatory funnel. Shake the funnel well, let it stand for a moment until the content in the funnel is separated into two distinct layers, and then collect the aqueous layer in an Erlenmeyer flask.

7. Wash the organic layer with saturated aqueous sodium chloride solution twice. Collect the organic layer in an Erlenmeyer flask and combine the aqueous layer with the reserved aqueous layer obtained via Step 6 after each washing.

8. Extract the combined aqueous layer with 15 mL of diethyl ether and combine the ether extract with the saved organic layer obtained via Step 7.

---

[24]　苯酚钠与溴乙烷的反应是非均相的。为了提高反应的效率，本实验采用了机械搅拌的方法，以促进反应物间的相互作用。

[25]　溴乙烷缓慢滴加，并控制反应温度在80～90 ℃，以避免反应过于剧烈，造成溴乙烷的逸出。

9. Transfer the combined organic layer to a dry and clean Erlenmeyer flask and dry the ether solution with anhydrous calcium chloride. Swirl the flask occasionally for 10-15 min until the ether solution is clear.

10. Decant the dried ether solution into a round-bottom flask carefully. Add several zeolites and assemble the flask for simple distillation to remove diethyl ether and other pre-fractions[26].

11. Assemble the apparatus for vacuum distillation to distill the final product and collect the pure ethyl phenyl ether in a pre-weighed receiving flask.

12. Weigh the final product and calculate the percentage yield of ethyl phenyl ether. Estimations of the boiling points of ethyl phenyl ether at different pressures are shown in Table 3.1.

Table 3.1 Estimations of the boiling points of ethyl phenyl ether at different pressures

Pressure/mmHg	5	10	20	40	60	100	200	400	760
Boiling point/℃	43	56	70	87	96	108	128	150	172

- **Helpful Hints**

1. Phenol is corrosive. Wear gloves and avoid skin contact. If phenol contacts with your skin, wash the affected area with copious amounts of cool water, and then scrub with a little ethanol.

2. Sodium hydroxide is also corrosive. Wear gloves and avoid skin contact with the solution of sodium hydroxide. If the basic solution contacts with your skin, wash the affected area with copious amounts of cool water.

3. Be careful to wear proper eye protection while performing vacuum distillation.

4. Phenol is a solid at room temperature (m.p. = 43 ℃). For easy access, you can put the reagent bottle containing phenol in a hot water-bath until phenol is melted, and then measure it with a cylinder[27].

5. The boiling point of ethyl bromide is 37-40 ℃. Control the reaction temperature at 80-90 ℃ so as to minimize the loss of ethyl bromide and minimize the side reactions.

6. The finial vacuum distillation to purify crude ethyl phenyl ether should be performed at moderate vacuum to avoid the loss of product[28].

- **Questions**

1. Why can Williamson ether synthesis not be used to prepare diisobutyl ether?

2. What is the purpose of each of the following operations in the work-up procedure of this experiment?

(1) Wash the organic layer with saturated aqueous sodium chloride solution twice.

---

[26] 减压蒸馏前，低沸点的溶剂和前馏分必须除去，以免对油泵造成损伤。
[27] 苯酚在室温下是固体（熔点为43 ℃），可以用热水浴温热使其熔化后量取。
[28] 最后一步的减压蒸馏不宜采用高真空，以免产物损失。

(2) Extract the aqueous layer with diethyl ether.

3. Why must diethyl ether and other low boiling point pre-fractions be removed prior to the final vacuum distillation?

4. Why is it important that equimolar amounts of sodium hydroxide and phenol are used in the Williamson ether synthesis of ethyl phenyl ether?

5. A student proposes to prepare t-butyl benzyl ether from t-butyl bromide and potassium benzyloxide. Predict the main product of the reaction.

6. Discuss the differences observed in the IR and ^{1}H NMR spectra of phenol and ethyl phenyl ether.

Exp.9　Preparation of o-tert-Butylhydroquinone

- **Objectives**

1. To study the principle and method of introducing alkyl group onto the aromatic rings by Friedel-Crafts alkylation.

2. To understand the importance of substituent effects for the electrophilic substitutions of aromatic rings.

3. To practice assembling the reflux apparatus with a mechanical stirrer.

4. To practice the washing and recrystallization of solid compounds.

- **Principle**

o-$tert$-Butylhydroquinone(TBHQ) is a new edible antioxidant that can be used to preserve oils, fats and food items, which can also be used as antioxidant in cosmetic products like lipsticks. o-$tert$-Butylhydroquinone(TBHQ) is usually prepared by the Friedel-Crafts alkylation reaction of hydroquinone with some alkylation agents such as isobutylene, $tert$-butyl alcohol or t-butyl methyl ether in the presence of acid[①].

In this experiment, o-$tert$-butylhydroquinone(TBHQ) is prepared by the reaction of hydroquinone with $tert$-butyl alcohol in xylene using phosphoric acid as the catalyst. The mechanism of this reaction includes two steps, i. e., the formation of intermediate of phenolic ether and the intermediate rearrangement to produce o-$tert$-butylhydroquinone. Herein, the second step is the rate-determining one, and a relatively long time will be required to accomplish this conversion at high temperature[②].

[①] 邻叔丁基对苯二酚可以通过对苯二酚的烷基化反应来合成。常用的烷基化试剂包括：异丁烯、叔丁醇和叔丁基甲基醚。

[②] 对苯二酚和叔丁基醇的烷基化反应一般分为两步：第一步是中间产物酚醚的生成，第二步是酚醚发生重排生成邻叔丁基对苯二酚。其中，第二步重排反应比较难以进行，需要在高温下反应较长的时间才能使中间产物充分转化。

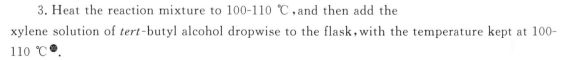

The alkylation of hydroquinone is a classic electrophilic aromatic substitution. *tert*-Butyl is an *ortho*- and *para*-directing activator, which makes benzene more reactive toward electrophilic reagent after the first *tert*-butyl is introduced onto the benzene ring. As a result, the second alkylation occurs easily and introduces the second *tert*-butyl onto the benzene ring. Considering the steric bulk of the *tert*-butyl group, the second substitution mainly occurs at the *p*-position of the first *tert*-butyl group and leads to the formation of 2,5-di-*tert*-butylhydroquinone[①]. Dropwise addition of the xylene solution of *tert*-butyl alcohol to hydroquinone is required during the reaction process to minimize the formation of this by-product.

Additionally, this reaction is heterogeneous and involves dropwise addition of the reactants. Therefore, vigorous stirring needs to be applied to ensure the complete interaction of the reactants, so as to facilitate the reaction[②].

- **Apparatus**

Apparatuses required include 100 mL three-necked flask, Claisen adapter, addition funnel, thermometer, Allihn condenser, Erlenmeyer flask, thermometer adapter, beaker, graduate cylinder, separatory funnel, Büchner funnel, filter flask, and apparatus for mechanical stirrer.

- **Setting Up**

Place 5.0 mL of 85% phosphoric acid and 20.0 mL of xylene into the 100 mL three-necked flask. Assemble the reaction apparatus as shown in Figure 3.11.

Figure 3.11 Apparatus for the preparation of *o-tert*-butyl hydroquinone

- **Experimental Procedures**

1. Turn on the stirrer, and add 5.5 g of hydroquinone into the reaction flask.

2. Place 7.7 mL of *tert*-butyl alcohol and 5.0 mL of xylene in the addition funnel.

3. Heat the reaction mixture to 100-110 ℃, and then add the xylene solution of *tert*-butyl alcohol dropwise to the flask, with the temperature kept at 100-110 ℃[③].

4. Upon the completion of the addition, heat the reaction mixture to reflux at 125-135 ℃

[①] 对苯二酚的烷基化是典型的亲电取代反应，叔丁基是邻、对位致活定位基团，苯环上上了一个叔丁基后，很容易再上一个叔丁基。由于位阻效应，第二个叔丁基主要进入原有叔丁基的对位，生成反应的主要副产物 2,5-二叔丁基对苯二酚。

[②] 由于此反应是非均相反应，且涉及反应物的滴加，因此反应需要在剧烈搅拌下进行。

[③] 叔丁基醇要缓慢地滴加到对苯二酚的甲苯溶液中，确保叔丁基醇的局部浓度不至于过高，以减少二取代产物的生成。

while stirring for another 1.5 h[⑭].

5. Stop heating and stirring, and pour the resulting mixture into a beaker containing 50 mL of hot water at once.

6. Cool down the solution until the solid is precipitated out.

7. Collect the white solids by vacuum filtration and wash the solids with 15 mL of water.

8. Transfer the filtrate into a separatory funnel and separate the waste mixture of xylene and phosphoric acid, and then dispose of them respectively[⑮].

9. Purify the crude *o-tert*-butylhydroquinone by recrystallization with 30 mL of xylene.

10. Collect the pure product by vacuum filtration.

11. Air-dry the product and calculate the percentage yield of *o-tert*-butyl hydroquinone.

• Helpful Hints

1. Phosphoric acid is corrosive. Wear gloves while handling it.

2. Hydroquinone is toxic and corrosive. Wear gloves and avoid skin contact while handling this reagent.

3. Xylene is a hazardous solvent and suspected carcinogen. Wear gloves and avoid skin contact while handling it. Perform this reaction in the hood to minimize exposure to it.

4. After the addition of *tert*-butyl alcohol, continuously heat the reaction mixture to reflux at higher temperature(125-135 ℃)to ensure the complete conversion of the intermediate phenolic ether to *o-tert*-butyl hydroquinone.

5. The xylene solution of *tert*-butyl alcohol should be added slowly to the reaction mixture to minimize the formation of di-*tert*-butylhydroquinone.

6. Using xylene as the reaction solvent can achieve two purposes: one is to reduce the local concentration of *tert*-butyl alcohol, thereby minimizing the formation of di-*tert*-butylhydroquinone, while the other is to remove the by-product of di-*tert*-butylhydroquinone due to its relatively large solubility in cold xylene[⑯].

• Questions

1. What are the commonly used alkylation reagents and acidic catalysts for Friedel-Crafts alkylation reaction?

2. What measures should be taken to improve the yield and purity of *o-tert*-butylhydroquinone in this experiment?

3. Why should the xylene solution of *tert*-butyl alcohol be added dropwise to the stirring reaction mixture in the initial step of this reaction?

4. In the work-up procedure, the reaction mixture is poured into hot water rather than

---

[⑭] 提高反应的温度是为了促进中间产物向产物的转化。

[⑮] 抽滤瓶中的滤液要经分液操作,将甲苯和磷酸分开,并按分类分别倒入不同的废液缸内。

[⑯] 本实验中二甲苯作为溶剂,有两个目的:一是控制叔丁醇浓度不至于过高,减少副产物二叔丁基对苯二酚的形成;二是副产物二叔丁基对苯二酚可以溶于冷的二甲苯溶液中,加入二甲苯可除去产物中的二叔丁基对苯二酚,对产物起到初步纯化的作用。

cold water. Explain the reason.

5. Discuss the differences observed in the IR and ^{1}H NMR spectra of hydroquinone and *o-tert*-butylhydroquinone.

3.4 Aldehydes and Ketones

Exp.10 Preparation of Cyclohexanone

- **Objectives**

1. To study the principle and method of synthesizing ketones by the oxidation of secondary alcohols using hypochlorous acid.

2. To practice assembling the reflux apparatus with electromagnetic stirrer.

3. To practice the washing and extraction of liquid compounds.

4. To learn to distill high boiling point liquids.

- **Principle**

One of the mostly used methods for preparation of ketones is the oxidation of secondary alcohols. Some common oxidizing agents such as potassium permanganate and chromic acid are extensively applied to the oxidation of secondary alcohols to form ketones[①]. Although potassium permanganate and chromic acid exhibit excellent oxidation ability, they are both derived from heavy metals hazardous to human health and environment. Moreover, the safe disposal of these oxidizing reagents and their reduction products is also troublesome and expensive.

In this experiment, cyclohexanone is synthesized by the oxidation of cyclohexanol, with sodium hypochlorite, a household bleach, taken as a green alternative oxidizing agent.

$$\text{C}_6\text{H}_{11}\text{OH} \xrightarrow[\text{CH}_3\text{COOH}]{\text{NaClO, H}_2\text{O}} \text{C}_6\text{H}_{10}\text{=O}$$

The mechanism by which sodium hypochlorite oxidizes cyclohexanol to form cyclohexanone probably involves the following steps:

Alkyl hypochlorite

The first is the reaction of cyclohexanol with hypochlorous acid, which is in equilibrium with hypohalite ion in aqueous medium, to form cyclohexyl hypochlorite. Then, base-promoted elimination of HCl directly leads to cyclohexanone. The green replacement of potassium permanganate or chromic acid with sodium hypochlorite for the oxidation of alcohols is endowed with some noticeable advantages, including mild reaction conditions, simple operation

---

① 仲醇的氧化是实验室合成酮常见的方法，常用的氧化剂为高锰酸钾或重铬酸。

and easy work-up procedure, etc[9]. In addition, this process is environmentally friendly[10], and it also saves the cost of waste disposal. The formed by-product chloride salt in this reaction can be safely flushed down the drain without further treatment.

- **Apparatus**

Apparatuses required include 250 mL three-necked flasks, 50 mL round-bottom flask, Allihn condenser, drying-tube containing sodium bicarbonate[11], Erlenmeyer flask, addition funnel, separatory funnel, distillation head, air condenser, thermometer, thermometer adapter, receiver adapter, graduate cylinder, beaker and apparatus for electromagnetic stirrer.

- **Setting Up**

Place 10.4 mL of cyclohexanol and 25 mL of acetic acid in a three-necked round-bottom flask. Assemble the reaction apparatus as shown in Figure 3.12?

Figure 3.12 Apparatus for the preparation of cyclohexanone

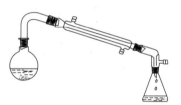

Figure 3.13 Apparatus for simple steam distillation

- **Experimental Procedures**

1. Place 80 mL of 11% sodium hypochlorite solution into the addition funnel.

2. Turn on the stirrer and add sodium hypochlorite solution dropwise to the stirring reaction mixture, with the reaction temperature kept within 35-40 ℃.

3. Continue to add sodium hypochlorite solution dropwise to the reaction mixture until the color of the solution in the flask turns into yellow-green.

4. Stop addition. Continuously stir the reaction mixture for 5 min and observe the color change of the solution. If the color of this solution turns colorless after stirring, continue to add sodium hypochlorite solution to the reaction mixture until the test of starch/iodide paper is positive[12].

5. Add another 5 mL of sodium hypochlorite solution and stir the reaction mixture for 15 min at room temperature.

---

⑨ 以次氯酸钠代替高锰酸钾或铬酸来氧化仲醇生成酮的方法，具有反应条件温和、操作简单、后处理方便等显著优点。

⑩ 环境友好的。

⑪ 为了吸收氧化过程中产生的氯气，可在冷凝管的上方加装装有固体碳酸氢钠的干燥管。

⑫ 可以用淀粉-碘化钾试纸来检测反应混合物中是否还有过量的次氯酸钠。

6. Add 1-4 mL of saturated sodium bisulfite solution dropwise to the reaction mixture until the test of starch/iodide paper becomes negative[12].

7. Stop heating and stirring. Add 50 mL of water and several zeolites to the flask when the mixture is no longer boiling, and assemble the flask for simple steam distillation[13] (Figure 3.13).

8. Distill the mixture and collect the distillate below 100 ℃.

9. Add some anhydrous sodium carbonate to the distillate in batches until no gas bubbles are observed. Then, add 10 g of sodium chloride to the mixture while stirring until all the sodium chlorides are completely dissolved[14].

10. Transfer the mixture to a separatory funnel and separate the layers. Save the organic layer in a dry Erlenmeyer flask.

11. Extract the water layer with 25 mL of *tert*-butyl methyl ether and combine the ether extract with the saved organic layer.

12. Dry the combined organic layer with anhydrous magnesium sulfate. Swirl the flask occasionally for 10-15 min until the liquid in the flask is clear.

13. Decant the dried ether solution into a dry round-bottom flask carefully. Add several zeolites, and assemble the flask for simple distillation to remove *tert*-butyl methyl ether.

14. Continue to heat the solution in the flask and collect the fraction boiling between 150-155 ℃ (760 mmHg) in a pre-weighed receiving flask[15].

15. Weigh the obtained product and calculate the percentage yield of cyclohexanone.

- **Helpful Hints**

1. Do not allow the solution of sodium hypochlorite to contact with your skin or eyes. If it does, flush the affected area immediately with copious amounts of water.

2. *tert*-Butyl methyl ether is volatile and irritating, so use it in a well-ventilated area and avoid prolonged contact or inhalation.

3. Sodium bicarbonate in drying-tube can absorb chlorine gas formed during the oxidation reaction.

4. Cyclohexanone is insoluble in water and can be distilled out with water at 95 ℃. The distillate collected in the step of simple steam distillation is a mixture of cyclohexanone, water and small amount of acetic acid.

5. The addition of saturated sodium bisulfite solution is to reduce excess sodium hypochlorite.

6. The test of starch/iodide paper is to determine the existence of excessive sodium hypochlorite.

---

[12] 加入饱和亚硫酸氢钠溶液的目的是除去过量的次氯酸钠。

[13] 此步的简易水蒸气蒸馏是为了蒸出粗产物环己酮（馏分中还含有水和少量的乙酸），以便于后面的提纯。

[14] 加入碳酸钠的目的是中和反应混合物中的乙酸。后面加入氯化钠可以降低环己酮在水中的溶解度，更便于分层。

[15] 环己酮的沸点为150~155 ℃（760 mmHg），所以此步的蒸馏要用空气冷凝管。

• **Questions**

1. Why is acetic acid employed in this reaction?

2. What is the function of saturated sodium bisulfite solution in the procedure of isolating cyclohexanone? Write down the related reaction equation.

3. What is the purpose of testing the reaction mixture with starch/iodide paper? Illustrate your answer by showing the chemical reaction that transforms iodide ions into iodine in this test.

4. The simple steam distillation is applied to distill out the crude cyclohexanone. Does the boiling point of the mixture ever exceed 100 ℃ during a steam distillation? Explain the reason.

5. Discuss the differences observed in the IR and ^1H NMR spectra of cyclohexanol and cyclohexanone?

Exp.11   Preparation of Acetophenone

• **Objectives**

1. To study the principle and method of preparing aryl ketones by Friedel-Crafts acylation reaction.

2. To learn to assemble the reflux apparatus with a gas-trap.

3. To practice the washing and extraction of liquid compounds.

4. To learn the technique of vacuum distillation.

• **Principle**

An acyl group can be introduced into an aromatic ring to provide an aryl ketone by the reaction of an arene with a carboxylic acid chloride(carboxylic acid anhydride) in the presence of aluminium trichloride. The overall reaction is known as the Friedel-Crafts acylation reaction⑯.

Friedel-Crafts acylation reaction is a typical electronohlic substituted reaction, where the reactive electrophile is a resonance-stabilized acyl cation generated by the reaction between the acyl chloride(acid anhydride) and aluminium trichloride⑰. The interaction of the vacant orbital on carbon with lone-pair electrons on the neighboring oxygen stabilizes the acyl ion. The stabilization prevents the occurrence of acyl cation rearrangement during acylation. Meanwhile, acylations hardly reoccur on an aromatic ring due to the deactivation of the introduced acyl group.

Compared with Friedel-Crafts alklation, Friedel-Crafts acylation has the advantages of a

---

⑯ 芳烃在路易斯酸存在条件下与酰氯或酸酐作用，芳环上的氢被酰基取代，生成芳香酮的反应称为付-克酰基化反应。

⑰ 在付-克酰基化反应中，酰氯或酸酐与路易斯酸作用生成稳定的酰基碳正离子，随后酰基碳作为亲电试剂进攻芳环，发生亲电取代反应生成芳香酮。

$$\left.\begin{array}{c}\underset{R}{\overset{O}{\underset{\|}{C}}}-Cl\\ \\ \underset{R}{\overset{O}{\underset{\|}{C}}}-O-\underset{R}{\overset{O}{\underset{\|}{C}}}\end{array}\right\}\xrightarrow{AlCl_3}\left[\begin{array}{c}\overset{\overset{..}{O}:}{\underset{R}{\overset{\|}{C+}}}\longleftrightarrow\overset{\overset{O^+}{\|}}{\underset{R}{\overset{\|}{C}}}\end{array}\right]\xrightarrow{Ar-H}\underset{\text{Acyl cation}}{\bigodot-\overset{O}{\underset{\|}{C}}-R}$$

high yield and pure product[⑱], making it widely applied in organic synthesis. In this experiment, acetophenone is synthesized by the acylation reaction of acetic anhydride with benzene in the presence of aluminum chloride.

$$\bigodot + (CH_3CO)_2O \xrightarrow{AlCl_3} \bigodot-COCH_3 + CH_3COOH$$

During the acetyl of benzene, acetophenone will form a complex with aluminum trichloride once it is formed. The finial product acetophenone is regenerated together with water-soluble aluminum salts when the complex is treated with dilute acid[⑲]. The whole process is shown as follows:

$$\bigodot-\overset{O:\ AlCl_3}{\underset{\|}{C}CH_3}\xrightarrow{H_3O^+}\bigodot-\overset{O}{\underset{\|}{C}CH_3} + Al(OH)Cl_2 + HCl$$

- **Apparatus**

Apparatuses required include 250 mL three-necked flask, 100 mL round-bottom flask, Allihn condenser, addition funnel, long-stem funnel, drying-tube containing anhydrous calcium chloride, separatory funnel, distillation head, Liebig condenser, thermometer, thermometer adapter, receiver adapter, Erlenmeyer flask, graduate cylinder, beaker, magnetic stirrer and apparatus for vacuum distillation.

- **Setting Up**

Add 30 mL of benzene into a dry 250 mL three-necked flask. Add several zeolites to the flask and assemble it for the reflux apparatus with a magnetic stirrer. Fit a drying tube containing anhydrous calcium chloride at the top of the condenser (Figure 3.14).

- **Experimental Procedures**

1. Weigh 20 g of anhydrous aluminum chloride and add it quickly into the reaction flask.
2. Place 6 mL of acetic anhydride and 10 mL of anhydrous benzene in the addition funnel, and swirl the solution gently.
3. Turn on the stirrer, and then add the benzene solution of acetic anhydride slowly to

---

[⑱] 因酰基碳正离子不会发生重排，同时酰基的钝化作用使取代反应能停留在一取代阶段，所以付-克酰基化反应具有收率高、产物纯的特点。

[⑲] 苯的乙酰化反应中生成苯乙酮和三氯化铝配合物，此配合物在无水介质中是稳定的，但在酸性条件下易水解，产物苯乙酮会重新析出。

the stirred reaction mixture. Control the addition rate and keep the solution slightly boiling. The addition will take 20-30 min[20].

4. Upon the completion of the addition, heat the reaction mixture to reflux gently in a water bath for an hour until no gas bubbles are released[21].

5. Stop heating and stirring. Cool the reaction mixture in an ice-water bath.

6. Place 50 mL of concentrated hydrochloric acid and 50 mL of ice water in the addition funnel, and then swirl the mixture gently.

7. Slowly add the dilute hydrochloric acid to the reaction mixture while continuously stirring in a vigorous manner until all the solids are dissolved[22].

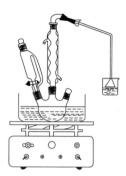

Figure 3.14　Apparatus for the preparation of acetophenone

8. Transfer the resulting solution into a separatory funnel, separate the layers and save the benzene layer in an Erlenmeyer flask.

9. Extract the water layer twice with 15 mL of benzene, and then combine the extracts with the saved benzene layer.

10. Wash the combined benzene layer sequentially with 20 mL of 5% sodium hydroxide solution and 20 mL of water.

11. Transfer the resulting benzene layer to a dry and clean Erlenmeyer flask, and dry it with a small amount of anhydrous magnesium sulfate. Swirl the flask occasionally for 10-15 min until the solution is clear.

12. Carefully transfer the solution to a clean and dry 100 mL round-bottom flask. Add several zeolites and assemble the flask for simple distillation to remove benzene and other pre-fractions at a low boiling point.

13. Assemble the apparatus for vacuum distillation, then distill the final product and collect the pure acetophenone at 86-90 ℃/12 mmHg in a pre-weighed dry receiving flask.

14. Weigh the final product and calculate the percentage yield of acetophenone.

• **Helpful Hints**

1. Anhydrous aluminum trichloride is extremely hygroscopic and reacts rapidly with water. To minimize exposure of this chemical to the atmosphere. Do not allow aluminum chloride to come in contact with your skin. If it does, flush the affected area with copious amounts of water.

2. To weigh and transfer anhydrous aluminum trichloride quickly to minimize deliquescence. The success of this experiment is highly dependent on the quality of the aluminum

---

⑳ 酰基化反应是放热的，乙酸酐的苯溶液应缓慢滴加，使苯慢慢回流，以免反应过于剧烈。

㉑ 随着乙酸酐的苯溶液的加入，三氯化铝逐渐溶解，氯化氢气体不断生成。待乙酸酐的苯溶液滴加完毕后反应平缓后，继续缓慢加热回流1h至无氯化氢气体放出为止。

㉒ 加完盐酸溶液后，若还有不溶物，可再补加适量稀盐酸至不溶物完全溶解。

chloride used[22].

3. Acetic anhydride is corrosive and irritating. Wear gloves and avoid skin contact while handling it. Perform this reaction in the hood to minimize exposure to it. Do not allow acetic anhydride to come in contact with your skin. If it does, flush the affected area with copious amounts of water.

4. Benzene is a hazardous solvent and suspected carcinogen. Wear gloves and avoid skin contact while handling it. Perform this reaction in the hood to minimize exposure to it.

5. All the chemicals and glassware used in the initial step of acylation should be dried completely[23].

6. The benzene solution of acetic anhydride should be added slowly to the reaction mixture. Otherwise, the acylation reaction will be hard to control.

7. The gas-trap system is used to absorb the hydrogen chloride formed during the acylation reaction. Place 200 mL of 20% sodium hydroxide solution in the beaker and take care to prevent back sucking[24].

8. The azeotropic boiling point of benzene and water is 69.4 ℃. A small amount of water in the crude acetophenone can be distilled out with benzene during simple distillation[25].

- **Questions**

1. Why is it important that the apparatus and reagents used for the Friedel-Crafts acylation should be dry?

2. In the acetylation of benzene, why is the dosage of the benzene excess?

3. What gas is responsible for the bubbles observed in the early stages of the reaction among benzene, acetic anhydride, and aluminum chloride?

4. Upon the completion of the initial acylation process, why is the dilute acid added to the reaction mixture?

5. An important difference between the Friedel-Crafts alkylation and acylation reactions is the dosage of Lewis acid. The latter process requires excess stoichiometric amount of aluminum chloride. Explain the reason based on the acetylation of benzene in this experiment.

Exp.12 Preparation of Benzaldehyde

- **Objectives**

1. To study the principle and method of preparing aldehydes by oxidation of primary alcohols.

2. To apply the phase transfer catalyst to facilitate a reaction.

3. To practice assembling the reflux apparatus with magnetic stirring.

---

[22] 三氯化铝的质量是实验成败的关键，其称量和投料都要迅速，不要将三氯化铝长时间暴露在空气中。

[23] 酰基化反应所用的仪器和药品必须绝对干燥。

[24] 本实验中的气体吸收装置是用来吸收反应中产生的氯化氢气体的。可在吸收装置的烧杯中加入 200 mL 20% 的氢氧化钠溶液，并注意调整漏斗的位置，以免倒吸。

[25] 苯与水可形成共沸物（b.p.=69.4 ℃）。因此，粗产物中少量的水在简单蒸馏时可以与苯一起蒸出。

4. To practice the washing and extraction of liquid compounds.

5. To learn the technique of vacuum distillation.

- **Principle**

One of the best methods of aldehyde synthesis is oxidation of primary alcohols. Primary alcohols can be oxidized to either aldehydes or carboxylic acids depending on the chosen reagents and the using conditions. Most commonly used oxidizing agents such as $KMnO_4$, CrO_3/CH_3COOH and $Na_2Cr_2O_7$ can oxidize primary alcohols directly to carboxylic acids. In the oxidation reactions above, an aldehyde is involved as an intermediate and will be further oxidized to a carboxylic acid rapidly. A common choice for preparing an aldehyde from a primary alcohol in the laboratory is to use pyridinium chlorochromate(PCC) in dichloromethane solvent[27]. Although pyridinium chlorochromate is a milder oxidizing agent with high selectivity, it is derived from heavy metal chromium, which is environmentally hazardous, thereby making the disposal of pyridinium chlorochromate and its reduction product troublesome and expensive. Fortunately, many new methods have been invented to oxidize primary alcohols to aldehydes using less hazardous reagents. A good example is optimizing the synthetic pathway towards benzaldehyde. In this experiment, benzaldehyde is to optimize prepared from benzyl alcohol using hydrogen peroxide as a green alternative oxidizing agent[28].

$$\text{PhCH}_2\text{OH} + H_2O_2 \xrightarrow[(C_4H_9)_4NHSO_4, 90^\circ C]{Na_2WO_4} \text{PhCHO} + H_2O$$

Being a mild and green oxidizing agent, hydrogen peroxide exhibits good selectivity for the oxidation of benzyl alcohol to benzaldehyde, while no further oxidation of benzaldehyde is observed. Benzyl alcohol usually features low solubility in the aqueous solution, which makes it necessary to apply a phase transfer catalyst known as tetrabutylammonium bisulfate to ensure the complete interaction among the reactants, and thus facilitate this heterogeneous reaction[29].

- **Apparatus**

Apparatuses required include 100 mL three-necked flask, 100 mL round-bottom flask, Allihn condenser, addition funnel, separatory funnel, distillation head, Liebig condenser, thermometer, thermometer adapter, receiver adapter, Erlenmeyer flask, graduate cylinder, beaker, magnetic stirrer and apparatus for vacuum distillation.

- **Setting Up**

Add 0.2 g of sodium tungstate, 0.2 g of tetrabutylammonium bisulfate, 7.5 mL of 30%

㉗ 氯铬酸的吡啶鎓盐（PCC）常在有机合成中作一级醇氧化成醛的氧化剂。此方法具有反应条件温和、选择性好、产物比较单一，且不容易发生过氧化等优点。

㉘ 本实验以双氧水为绿色替代传统氧化剂，在水溶液中氧化苯甲醇来制备苯甲醛。此方法不仅减少了对人体和环境的伤害，同时也降低了反应的成本。

㉙ 双氧水氧化苯甲醇的反应是非均相的。为了保证反应物之间的充分作用，促进反应的进行，反应过程中需添加相转移催化剂。

hydrogen peroxide and 10 mL of water in a 100 mL three-necked round-bottom flask. Assemble the reaction apparatus as shown in Figure 3.15.

- **Experimental Procedures**

1. Place 6.5 g of benzyl alcohol into the addition funnel.

2. Turn on the stirrer, and add benzyl alcohol dropwise to the stirred reaction mixture in the flask.

3. Heat the mixture while stirring for 3 h, with the reaction temperature kept at about 90 ℃.

Figure 3.15 Apparatus for the preparation of benzaldehyde

4. Add saturated sodium thiosulfate solution dropwise to the resulting mixture from the addition funnel until the test of starch/iodide paper becomes negative[30].

5. Stop heating and stirring. When the mixture is no longer boiling, add several zeolites to the flask again and assemble the flask for simple steam distillation[31].

6. Distill the mixture and collect the distillate until the distillate appears to be clear.

7. Transfer the distillate to a separatory funnel. Shake the funnel well and let it stand for a moment until the content in the funnel is separated into two distinct layers. Then, save the organic layer in an Erlenmeyer flask.

8. Extract the water layer with *tert*-butyl methyl ether and combine the ether extract with the saved organic layer.

9. Wash the combined organic layer with saturated sodium thiosulfate and remove the water layer.

10. Transfer the organic layer to a dry and clean Erlenmeyer flask, and dry it with anhydrous magnesium sulfate. Swirl the flask occasionally for a period of 10~15 min until the liquid in the flask is clear.

11. Carefully decant the dried ether solution into a round-bottom flask. Add several zeolites and assemble the flask for simple distillation to remove *tert*-butyl methyl ether.

12. Assemble the apparatus for vacuum distillation, then distill the final product and collect the pure benzaldehyde at 59-61 ℃/10 mmHg in a pre-weighed receiving flask[32].

13. Weigh the final product and calculate the percentage yield of benzaldehyde.

- **Helpful Hints**

1. *tert*-Butyl methyl ether is volatile and irritating. Use it in a well-ventilated area and avoid prolonged contact or inhalation.

2. Be careful to wear proper eye protection while performing vacuum distillation.

3. The purpose of simple steam distillation is to distill out the crude benzaldehyde from

---

[30] 用饱和硫代硫酸钠溶液洗涤是为了除去反应体系中过量的氧化剂过氧化氢,以免生成的苯甲醛继续氧化。过氧化氢是否完全除去,可用淀粉-碘化钾试纸进行检测。

[31] 简易水蒸气蒸馏的目的是从反应混合物中蒸出粗产物苯甲醛,以便于后面的分离、提纯。

[32] 苯甲醛的沸点是179 ℃,但苯甲醛长时间受热易被氧化生成苯甲酸。所以,苯甲醛的提纯要采用减压蒸馏。

the reaction mixture to facilitate the following work-up procedure.

4. The saturated sodium thiosulfate solution is used to remove excess hydrogen peroxide and avoid the further oxidation of benzaldehyde.

5. The boiling point of benzaldehyde is 179 ℃. Benzaldehyde is easily oxidized to form benzoic acid when heated for a long time. Thus, vacuum distillation is applied to purify benzaldehyde finally.

6. *tert*-Butyl methyl ether is a low-boiling point solvent, and should be removed by simple distillation prior to the final vacuum distillation to avoid damage to the oil pump[133].

- **Questions**

1. In this experiment, what measures should be taken to remove the reacted starting material benzyl alcohol?

2. What is the function of sodium thiosulfate in the procedure of benzaldehyde isolation? Write down the related reaction equations.

3. What is the purpose of testing the reaction mixture with starch/iodide paper? Illustrate your answer by showing the corresponding chemical reaction.

4. Briefly describe the features of phase transfer catalysts. What other phase transfer catalysts can be applied in this experiment?

5. Please describe the principle of steam distillation.

6. Discuss the differences observed in the IR and ^{1}H NMR spectra of benzaldehyde and benzyl alcohol.

3.5 Carboxylic Acids

Exp.13　Preparation of Benzoic Acid

- **Objectives**

1. To study the principle and method of preparing aromatic acids from aromatic hydrocarbons.

2. To practice assembling the reflux apparatus with mechanical stirring.

3. To practice precipitation by pH adjustment.

4. To practice the washing and recrystallization of solid compounds.

- **Principle**

Aromatic carboxylic acids can be synthesized by the oxidation of aromatic hydrocarbons.

Aromatic rings are too inert to be oxidized by common oxidizing agents, such as potassium permanganate and potassium dichromate. However, the alkyl side chains that are attached to an aromatic ring, and are dramatically affected by the aromatic ring, will react rapidly with oxidizing agents and can be converted into carboxyl group, —COOH[134]. The net effect of this

---

[133] 甲基叔丁基醚是常用的萃取剂，其沸点为 55 ℃。在减压蒸馏前，应先通过常压蒸馏回收。

[134] 芳环上的侧链受苯环的影响，很容易被氧化剂氧化生成羧基。

oxidation is the conversion of an alkylbenzene into a benzoic acid. For instance, methyl benzene can be oxidized by aqueous $KMnO_4$ to give benzoic acid. In this oxidation process, purple potassium permanganate is reduced to water-insoluble brown manganese dioxide, while methyl benzene is oxidized to water-soluble sodium benzoate. After filtering out manganese dioxide, benzoic acid can be recovered by acidifying the resulting aqueous solution of benzoate[⑱].

$$C_6H_5CH_3 + 2KMnO_4 \longrightarrow C_6H_5COOK + 2MnO_2 + KOH + H_2O$$

$$C_6H_5COOK + HCl \longrightarrow C_6H_5COOH + KCl$$

The mechanism of side-chain oxidation is complex, and may involve the intermediate benzylic radicals formed by the cleavage of the C—H bonds next to the aromatic ring. Therefore, as long as a side chain contains benzylic hydrogen, it, no matter how long it is, will be eventually oxidized to a carboxyl group. Conversely, *tert*-butylbenzene has no benzylic hydrogens, and is therefore inert[⑲].

Side-chain oxidation reactions are exothermic and vigorous, making it crucial to control the reaction temperature to minimize side reactions and explosion risks.

- **Apparatus**

Apparatuses required include 250 mL three-necked flask, Allihn condenser, graduate cylinder, beaker, Büchner funnel, filter flask, and apparatus for mechanical stirrer.

- **Setting Up**

Sequentially add 8.5 g of potassium permanganate, 2.7 mL of toluene and 100 mL of water into the three-necked flask. Then, assemble the apparatus as shown in Figure 3.16.

Figure 3.16 Apparatus for the preparation of benzoic acid

- **Experimental Procedures**

1. Heat the reaction mixture to reflux under stirring[⑳] for 2 h until the oily layer disappears.

2. Stop heating, and then add about 5 mL of saturated sodium bisulfite solution through the condenser to the resulting mixture while stirring.

3. Stop stirring when the mixture is no longer purple.

---

⑱ 高锰酸钾氧化甲苯制备苯甲酸的反应过程中，高锰酸钾被还原生成不溶于水的二氧化锰，甲苯被氧化生成水溶性的苯甲酸的钾盐。过滤除去二氧化锰，用盐酸酸化苯甲酸钾的水溶液即可得到苯甲酸。

⑲ 芳烃的侧链不管有多长，只要含有α-氢，最终都会被氧化生成羧基。叔丁基苯不含有α-氢，因此不会被高锰酸钾氧化。

⑳ 此反应是非均相反应，为了使反应物能充分接触，避免局部过浓、过热而导致其他副反应或反应物的分解，反应物需进行强烈搅拌。

4. Filter the mixture under vacuum when it is hot and collect the filtrate in a beaker[38].

5. Wash the filter cake with 10 mL of hot water, and then combine the washing liquid to the reserved filtrate.

6. Allow the combined filter to be cooled down in an ice-water bath.

7. Slowly add 5-10 mL of concentrated hydrochloric acid dropwise to the cooled solution while stirring until the solution is slightly acidic (pH=3-5)[39].

8. Collect the solid by vacuum filtration, and then wash the crude product with 5 mL of cold water[40].

9. Dry the product under infrared light.

10. Weigh the dried product and calculate the percentage yield of benzoic acid.

11. Benzoic acid can be further purified by being recrystallized from hot water.

- **Helpful Hints**

1. Toluene is a hazardous solvent and suspected carcinogen. Wear gloves and avoid skin contact while handling it. Perform this reaction in the hood to minimize exposure to it.

2. Potassium permanganate is a strong oxidant. Avoid contact as it will stain skin and clothing. Wash the area thoroughly with soap and water if affected.

3. Concentrated hydrochloric acid fume is suffocative and corrosive, so handle it in the hood. If any acid spills on your skin, wash it off with large amount of water and then with dilute sodium bicarbonate solution.

4. Try to keep the reaction temperature not too high and maintain the liquid refluxing gently in the oxidation process[41].

5. Hot filtration is required to remove the formed manganese dioxide. Do not forget to preheat the Büchner funnel and filter flask for filtration.

6. Upon the completion of the oxidation reaction, saturated sodium bisulfite solution is added to the mixture to reduce excess potassium permanganate.

7. Keep stirring during the oxidation reaction to avoid bumping and spraying[42].

- **Questions**

1. Why must the mechanical stirrer be applied in the oxidation reaction of toluene to benzoic acid when potassium permanganate is used as the oxidizing agent?

2. How should the end point of the above reaction be determined?

3. Can saturated sodium bisulfite be used to replace saturated sodium thiosulfate in Step 2 of the experimental procedures? Explain the reason.

4. If the toluene has not been fully oxidized to benzoic acid, what measures should be

---

[38] 趁热过滤时，布氏漏斗和抽滤瓶应事先预热。过滤后如果滤液仍呈紫色，可继续加热反应一段时间或用饱和亚硫酸氢钠溶液还原至无色。

[39] 随着盐酸的加入，会有白色固体析出。酸化反应是放热的，盐酸要分批、边搅拌边加入。

[40] 苯甲酸在水中有一定的溶解度，洗涤时要用少量冷水来洗。

[41] 反应的温度不宜太高，保持回流状态即可。

[42] 在整个氧化反应进行的过程中，不要停止搅拌，以防止暴沸和喷液。停止反应时，要先关掉电源并撤去电热套，继续搅拌5～10min，再停止搅拌。

taken to remove the unreacted toluene?

5. Are there any other methods for the preparation of benzoic acid? Write down the corresponding reaction equations.

6. Discuss the differences observed in the IR and ^{1}H NMR spectra of toluene and benzoic acid.

Exp.14   Preparation of Cinnamic Acid

- **Objectives**

1. To study the principle and method of preparing α,β-unsaturated acid by Perkin reaction.

2. To practice assembling the reflux apparatus with an air-condenser.

3. To learn the technique of steam distillation.

4. To practice precipitation by pH adjustment.

5. To learn the technique of melting-point measurement with a digital melting-point apparatus.

- **Principle**

Aldol condensation reaction takes place between two molecules of aldehydes/ketones in the presence of bases or acids, which involves a combination of an α-substitution and a nucleophilic addition. In this reaction, one molecule of aldehyde or ketone containing α-H is converted into an enolate-ion nucleophile (α-substitution) and is added to the electrophilic carbonyl group of the second molecule of aldehyde or ketone (nucleophilic addition) to yield β-hydroxyl aldehydes or β-hydroxyl ketones, which can quickly undergo dehydration under acidic condition or upon heating to generate the final product α,β-unsaturated aldehydes or ketones. An aldol reaction combines two molecules of reactant by forming a new carbon-carbon bond, and thus provides a good method to build larger molecules from smaller precursors[⑱].

Perkin condensation is a typical aldol-type condensation, where an aromatic aldehyde (ArCHO) reacts with a carboxylic acid anhydride containing α-H with salts taken as the base catalyst, and carboxylate salts and carbonate are usually used.

$$\text{ArCHO} + (\text{RCH}_2\text{CO})_2\text{O} \xrightarrow[\triangle]{\text{Alkali salts}} \text{ArCH}=\text{CRCOOH} + \text{RCH}_2\text{COOH}$$

During this process, a carboxylic acid anhydride with α-H is converted into its enolate ion by the base catalyst. The enolate ion acts as a nucleophilic donor and and is added to the carbonyl group of an aromatic aldehyde with no α-H. Then, a mixed aldol condensation reaction between an aromatic aldehyde and a carboxylic acid anhydride will be successfully completed and the final product of β-aryl-α,β-unsaturated carboxylic acid (ArCH=CRCOOH)

---

⑱ 在羟醛缩合反应中,一分子含有 α-H 的醛或酮分子在稀碱催化下转化为具有亲核性的烯醇负离子,接着与第二分子醛或酮分子中的亲电性的羰基发生亲核加成,生成 β-羟基醛或酮。β-羟基醛或酮受热脱水,生成最终产物 α、β-不饱和醛或酮。通过羟醛缩合反应,可以在分子中形成新的碳碳键,并增长碳链。

will be generated[44]. For instance, cinnamic acid can be synthesized from benzaldehyde and acetic anhydride by Perkin reaction using potassium acetate as a base catalyst, and the reaction is indicated as below:

$$\text{PhCHO} + (\text{CH}_3\text{CO})_2\text{O} \xrightarrow[\Delta]{\text{KAc}} \text{PhCH}=\text{CHCOOH} + \text{CH}_3\text{COOH}$$

The Perkin reaction between benzaldehyde and acetic anhydride needs to be performed at high temperature, yet the main reaction may be accompanied by some side reactions, such as decarboxylation and polymerization if the reaction mixture is over heated. In this case, considerable emphasis should be attached to controlling the reaction temperature between 150-170 ℃[45].

$$\text{PhCH}=\text{CHCOOH} \xrightarrow{\Delta} \text{PhCH}=\text{CH}_2$$

$$n\,\text{PhCH}=\text{CH}_2 \longrightarrow \left[\underset{\text{Ph}}{\overset{\text{H}}{\text{C}}}-\text{CH}_2\right]_n$$

● **Apparatus**

Apparatuses required include 150 mL three-necked round-bottom flask, air condenser, 250 round-bottom flask, thermometer, thermometer adapter, graduate cylinder, 250 mL beaker, Büchner funnel, filter flask and apparatus for steam distillation.

● **Setting Up**

Place 3.0 g of potassium acetate, 3.0 mL of benzaldehyde, 5.5 mL of acetic anhydride and several zeolites into a 150 mL three-necked round-bottom flask. Assemble the apparatus as shown in Figure 3.17.

● **Experimental Procedures**

1. Heat the reaction mixture to reflux for 1 h, with the reaction temperature kept between 150-170 ℃.

2. Cool down the reaction mixture to about 100 ℃, and then add 20 mL of hot water into the reaction flask.

3. Add 15-20 mL of saturated sodium carbonate into the flask while stirring until the mixture is slightly basic(pH=8)[46].

---

[44] Perkin 反应是指含有 α-H 的酸酐在碱性催化剂的作用下,转化成为相应的烯醇负离子,接着烯醇负离子与不含有 α-H 的芳香醛发生亲核加成生成 β-芳基-α,β-不饱和羧酸的反应。Perkin 反应常用的碱性催化剂为与酸酐对应的羧酸的钠盐、钾盐,也可用碳酸钾等代替。

[45] Perkin 反应需要在较高的温度下完成,但反应温度也不宜太高,以免缩合产物在高温下发生脱羧、聚合等副反应。

[46] 加入饱和碳酸钠溶液的目的是中和反应生成的副产物乙酸,同时将产物肉桂酸全部转化成相应的钠盐溶解在水中,以免影响后面的水蒸气蒸馏。

4. Assemble the reaction flask for steam distillation (Figure 3.18) and distill the resulting mixture until the distillate appears to be clear⑰.

Figure 3.17 Apparatus for the preparation of cinnamic acid

Figure 3.18 Apparatus for steam distillation

5. Transfer the residue in the flask into a beaker, and then add a small amount of water until the total amount of liquid is 125 mL.

6. Add about 1.0 g of activated carbon into the solution and heat the solution to boil for 5 min⑱.

7. Separate the solid and liquid by hot vacuum filtration.

8. Transfer the filtrate into a clean beaker and slowly add 10-20 mL of concentrated hydrochloric acid dropwise to the solution until the Congo red test paper turns red (pH=3).

9. Allow the solution to be cooled down in a cold-water bath to completely precipitate the product.

10. Filter the solution and wash the solid with 5-10 mL of cold water.

11. Air-dry the product and calculate the percentage yield of cinnamic acid.

12. Cinnamic acid can be purified by recrystallization with a mixed solvent of water and ethanol with the volume ratio of 3∶1.

13. Measure the melting point of cinnamic acid.

• **Helpful Hints**

1. Acetic anhydride is corrosive and irritating. Wear gloves and avoid skin contact while handling it. Perform this reaction in the hood to minimize exposure to it. Do not allow acetic anhydride to come in contact with your skin. If it does, flush the affected area with copious amounts of water.

2. Benzaldehyde is toxic and irritating. Wear gloves and avoid skin contact while handling it.

3. Benzaldehyde is easily oxidized to form benzoic acid and should be redistilled

---

⑰ 水蒸气蒸馏的目的是除去没有反应完的苯甲醛,起到初步纯化的作用。

⑱ 反应物长时间加热,会发生部分脱羧生成不饱和的烃类副产物,并进而生成树脂状物。此处加入活性炭不仅是为了脱色,同时活性炭也可以吸附生成的树脂状杂质。

before use[19].

4. Acetic anhydride may be converted into acetic acid when exposed to moisture for a long time, and should therefore be redistilled before use as well.

5. Some bubbles will appear in the initial stage of the reaction due to the evolution of carbon dioxide.

6. It is crucial to control the reaction temperature between 150-170 ℃, which not only ensures the reaction to proceed, but also minimizes the side reactions.

7. Cinnamic acid has two *cis-* and *trans-* isomers, and the product prepared by Perkin reaction is usually *trans*-isomer with a melting point of 135.6 ℃[20]. *trans*-Cinnamic acid is an irritant. Wear gloves and avoid skin contact with it.

8. Cinnamic acid may undergo polymerization.

• **Questions**

1. What measures should be taken to minimize the side reactions in this experiment?

2. What can be obtained from the Perkin reaction of benzaldehyde and propanoic anhydride in the presence of anhydrous potassium carbonate? Write down the corresponding reaction equation.

3. What is the principle of steam distillation? Why is steam distillation applied in this experiment?

4. Saturated sodium carbonate solution is used to adjust the pH of the mixture to 8 prior to steam distillation. What undesired reaction might occur if a concentrated solution of sodium hydroxidethe is used to replace sodium carbonate?

5. Give the structures of *trans-* and *cis-* isomers of cinnamic acid. What methods can be used to distinguish the two kinds of isomers?

Exp.15 Preparation of Hexane-1,6-dionic Acid

• **Objectives**

1. To study the principle and method of preparing hexane-1,6-dionic acid from cyclohexene.

2. To practice assembling the reflux apparatus with mechanical stirrer.

3. To practice precipitation by pH adjustment and vacuum filtration.

4. To learn the concentration of a liquid.

5. To practice recrystallization.

• **Principle**

Oxidation reaction is the most commonly used method to prepare carboxylic acids. Carboxylic acids can be synthesized by oxidation of olefins, primary alcohols, or aldehydes. Several oxidizing reagents are involved in these reactions, including potassium permanganate, po-

---

[19] 苯甲醛使用前需重新蒸馏，否则产物中可能会含有苯甲酸等杂质，很难分离。

[20] 肉桂酸有顺式和反式两种异构体。实验中得到的产物是反式肉桂酸，其熔点为135.6 ℃。

tassium dichromate, nitric acid, and H_2O_2, etc.

As a typical strong oxidizing reagent, potassium permanganate($KMnO_4$) can oxidize olefins to carbonyl-containing products with the cleavage of the double bond of the olefines. The permanganate oxidation of olefines can afford ketones, carboxylic acids or carbon dioxide depending solely on the structure of the olefins[51]. For example, two ketone fragments are formed when an alkene with a tetrasubstituted double bond is oxidized by $KMnO_4$, one ketone and one caboxylic acid is formed when an alkene with a trisubstituted double bond is oxidized, and carbon dioxide is formed when two hydrogens are present on one double bond carbon.

$$\underset{R'}{\overset{R}{>}}C=C\underset{R_2}{\overset{R_1}{<}} \xrightarrow[H_2O]{KMnO_4} \underset{R'}{\overset{R}{>}}C=O + O=C\underset{R_2}{\overset{R_1}{<}}$$

$$\underset{R'}{\overset{R}{>}}C=CHR_1 \xrightarrow[H_2O]{KMnO_4} \underset{R'}{\overset{R}{>}}C=O + R_1-\overset{O}{\underset{\|}{C}}-OH$$

$$\underset{R'}{\overset{R}{>}}C=CH_2 \xrightarrow[H_2O]{KMnO_4} \underset{R'}{\overset{R}{>}}C=O + HO-\overset{O}{\underset{\|}{C}}-OH \downarrow CO_2+H_2O$$

In this case, when a cyclic olefin serves as the substrate, two carbonyl groups (either ketone carbonyl or carboxyl group) are generated at the termini of the same molecule. As a result, the oxidation of cyclic olefin provides a good method to introduce two functional groups to the molecule at same time[52]. In this experiment, cyclohexene with no substituent on the double bond carbons is oxidized to give hexane-1,6-dionic acid (adipic acid).

$$\text{cyclohexene} \xrightarrow[H_2O]{KMnO_4} \text{cyclohexane-1,2-dicarboxylic (COOH, COOH)}$$

• **Apparatus**

Apparatuses required include 250 mL three-necked flask, Allihn condenser, thermometer, thermometer adapter, graduate cylinder, 250 mL beaker, Büchner funnel, filter flask and apparatus for mechanical stirrer.

• **Setting Up**

Add 8.4 g of potassium permanganate, 2.0 mL of cyclohexene and 50 mL of water into a three-necked flask sequentially, and then, assemble the apparatus as shown in Figure 3.19.

---

[51] 烯烃在酸性高锰酸钾作用下，双键断开，氧化生成含有羰基的化合物。根据烯烃结构的不同，高锰酸钾氧化烯烃的最终产物分别为酮、羧酸或二氧化碳。

[52] 当环状烯烃作为反应底物时，随着双键的断开，会在同一分子的两端产生两个羰基（酮羰基或羧基）。因而，环状烯烃的高锰酸钾氧化为在同一分子中同时引入两个官能团提供了一种很好的方法。

• **Experimental Procedures**

1. Heat the mixture in the flask over a 45 ℃ water bath for 30 min under vigorous stirring. Then, heat the resulting solution to reflux for 15 min[⑱].

2. Slowly add about 1 mL of methanol to the reaction mixture while stirring until the reaction solution turns into colorless[⑲].

3. Allow the reaction mixture to be cooled down to room temperature and separate the solid and liquid by vacuum filtration.

4. Rinse the reaction flask with 2×10 mL of hot 1% sodium hydroxide solution and pour the washing solution to wash the filter cake.

5. Transfer the combined filtrate to a 250 mL beaker, and heat the solution to boil while stirring until the volume of the solution is about 10 mL.

6. Cool down the resulting solution in an ice-water bath.

7. Slowly and carefully add concentrated hydrochloric acid to the cooled solution under vigorous stirring until the solution is acidic (pH=1).

8. Stand the beaker in an ice-water bath for 10 min to precipitate the product completely.

9. Collect the crude hexane-1,6-dionic acid by vacuum filtration.

10. Purify the crude product by recrystallization with no more than 10 mL of hot water.

11. Weigh the product and calculate the percentage yield of hexane-1,6-dionic acid.

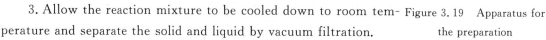

Figure 3.19 Apparatus for the preparation of hexane-1,6-dionic acid

• **Helpful Hints**

1. Cyclohexene is toxic and irritating. Wear gloves and avoid skin contact while handling it. Perform this reaction in the fume hood to minimize exposure to it.

2. Potassium permanganate is a strong oxidant. Avoid contact as it will stain skin and clothing.

3. Concentrated hydrochloric acid fume is suffocative and corrosive, so handle it in the fume hood. If any acid spills on your skin, wash it off with large amount of water and then with dilute sodium bicarbonate solution.

4. The oxidation of cyclohexene is exothemic, so control the reaction temperature to avoid the bumping and spraying of the reaction mixture.

5. One can check whether the reaction mixture is colorless after methanol is added in the following way: take a stir rod and tip a little of reaction mixture onto a filter paper, and then observe whether a purple circle or a colorless circle appears around the dark brown manganese dioxide.

---

⑱ 氧化反应是放热的，控制反应的温度不要太高，以防止反应过于激烈而引起反应液的暴沸。

⑲ 随着反应的进行，高锰酸钾被还原生成黑褐色的二氧化锰。加入甲醇后，很难准确判定反应液是否由紫色变为无色。此时，可以用玻璃棒蘸取少量反应混合物沾在滤纸上，然后观察黑褐色二氧化锰周围出现圆圈的颜色。

6. Concentrated hydrochloric acid should be added carefully in the fume hood[⑮].

7. The solubility of adipic acid in water varies greatly with temperature. Therefore, the resulting solution after adding concentrated hydrochloric acid should be cooled down in an ice-water bath for complete crystallization[⑯].

- **Questions**

1. Upon the completion of the oxidation, why is a small amount of methanol added to the reaction mixture? Write down the corresponding reaction equations.

2. Why is the concentration of the solution necessary before it is acidified by concentrated hydrochloric acid?

3. Are there any other methods for the preparation of hexane-1,6-dionic acid? Write down the corresponding reaction equations.

4. Discuss the differences observed in the IR and ^{1}H NMR spectra of cyclohexene and hexane-1,6-dionic acid.

3.6 Carboxylic Acid Derivatives

Exp.16 Preparation of Ethyl Acetate

- **Objectives**

1. To study the principle and method of preparing esters by Fischer esterification.

2. To practice setting up a reaction by refluxing.

3. To practice filtration, washing, drying and distillation.

- **Principle**

Esters can be synthesized by the nucleophilic acyl substitution reaction of a carboxylic acid with alcohol, i.e., the Fischer esterification reaction. Discovered by the Nobel laureate Emil Fischer, Fischer esterification is one of the classic reactions in organic chemistry, usually catalyzed by a mineral acid, such as concentrated sulfuric acid or hydrochloric acid. Herein, the acid catalyst protonates carbonyl-group oxygen atoms, giving the carboxylic acid a positive charge and making it much more electrophilic, thereby effectively promoting the nucleophilic addition of a weakly nucleophilic alcohol to a carboxylic acid[⑰]. The preparation of ethyl acetate is a good example of Fischer esterification.

$$CH_3COOH + C_2H_5OH \underset{Refluxing}{\overset{H_2SO_4}{\rightleftharpoons}} CH_3COOC_2H_5 + H_2O$$

Fischer esterification is an equilibrium reaction, in which acid-catalyzed hydrolysis of an

---

⑮ 浓盐酸的酸化要在通风橱中进行，盐酸应边搅拌边缓慢滴加。

⑯ 己二酸微溶于水中，但其在水中的溶解度随温度变化很大。因此，加入浓盐酸后的溶液应放在冰水浴中冷却，以确保产物结晶完全。

⑰ 在费舍尔酯化反应中，酸性催化剂通过质子化羰基氧来活化羧酸中的羰基，增强其亲电性，从而促进弱亲核性的醇与羧酸的亲核加成。

ester is the reverse of esterification. This equilibrium is readily established upon heating. In order to push this equilibrium to the right and increase the yield of the produced ethyl acetate, inexpensive reactant ethanol is used in this conversion. Meanwhile, acid catalyst concentrated sulfuric acid can also serve as the dehydrating reagent to remove the by-product water, so as to drive the reaction forward[58].

Some side-reactions may readily occur with the main reaction, which are shown as follows:

$$2C_2H_5OH \xrightleftharpoons[130\sim150\ ℃]{H_2SO_4} C_2H_5OC_2H_5 + H_2O$$

$$2C_2H_5OH \xrightleftharpoons[160\sim180\ ℃]{H_2SO_4} CH_2=CH_2 + H_2O$$

$$C_2H_5OH \xrightarrow[>180\ ℃]{H_2SO_4} CO_2 + C + H_2O$$

These side reactions can be minimized by controlling the esterification temperature within the suitable range.

- **Apparatus**

Apparatuses required include 50 mL round-bottom flask, Allihn condenser, thermometer, thermometer adapter, distillation head, Liebig condenser, adapter, Erlenmeyer flask, separatory funnel and graduate cylinder.

- **Setting Up**

Place 14.5 mL of anhydrous ethanol, 9.0 mL of acetic acid and several zeolites into a dry 50 mL round-bottom flask, and then add 3.8 mL of concentrated sulfuric acid dropwise to the mixture carefully[59]. Mix the reaction mixture thoroughly by swirling the flask and assemble the reaction apparatus as shown in Figure 3.20.

Figure 3.20　Apparatus for the preparation of ethyl acetate

- **Experimental Procedures**

1. Slowly heat the reaction mixture to reflux gently for 0.5 h[60].

2. Stop heating and cool down the reaction mixture to room temperature.

3. Add several zeolites to the flask again and assemble the reaction flask for simple distillation.

4. Collect the distillate with a flask immersed in a cold-water bath[61].

5. Terminate distillation when the distillate is about half of the total volume of react-

---

⑤⑧　酯化反应中，酸不仅是催化剂，同时也可以作为脱水剂来除去反应中生成的副产物水，使平衡反应右移，提高酯的收率。

⑤⑨　硫酸应分批滴加，且要边加边摇匀，防止局部浓度过高受热炭化。

⑥⓪　反应温度的控制很重要。温度过高，副反应增多，且易造成反应物和产物的逸出。温度过低，反应速度慢，产率低。

⑥①　此步蒸馏的目的是蒸出粗产物乙酸乙酯，便于后面的提纯。同时随着乙酸乙酯的蒸出，反应也会进一步右移，促使反应完全。蒸馏时速度不能过快，以免反应物在高温条件下发生炭化。

ants.

6. Slowly add saturated aqueous sodium carbonate into the collected distillate while stirring until carbon dioxide gas is no longer released and the solution is brought to a pH value of about 8.

7. Transfer the solution to a separatory funnel and discard the aqueous layer.

8. Wash the organic layer sequentially with 5.0 mL of saturated aqueous sodium chloride solution, 5.0 mL of saturated calcium chloride, and 5.0 mL of water[①].

9. Pour the organic layer into a clean and pre-dried Erlenmeyer flask and add 0.3 g of anhydrous magnesium sulfate to dry the solution for 15-20 min.

10. Carefully decant the organic layer into a pre-dried 50 mL round-bottom flask and assemble the flask for simple distillation.

11. Collect the fraction at 73-78 ℃ (760 mmHg) in a pre-weighed dry receiving flask.

12. Weigh the product and calculate the percentage yield of ethyl acetate.

- **Helpful Hints**

1. Concentrated sulfuric acid is corrosive and can cause severe burns. Be extremely careful while handling it. If any concentrated sulfuric acid comes in contact with your skin, immediately wash it off with copious amounts of cold water and then with dilute sodium bicarbonate solution.

2. Acetic acid is corrosive and suffocate, so handle it in the hood and wear gloves. If any acid spills on your skin, wash it off with large amount of water.

3. After concentrated sulfuric acid is added dropwise to the flask, the reaction mixture should be mixed well to prevent carbonization of the reactants.

4. Make sure that all the joints of the ground glassware are sealed well during refluxing and distillation to prevent the reactants and products from escaping.

5. Slowly heat the reaction mixture to reflux to minimize the side reactions.

6. There are two purposes of using saturated aqueous sodium chloride solution instead of water to wash the organic layer: one is to prevent emulsification and aid in layer separation, and the other is to reduce the loss of ethyl acetate due to its solubility in water[②].

7. Ethyl acetate can form azeotrope with water(b. p. =70.4 ℃)or with ethanol(b. p. =70.8 ℃). A ternary azeotrope(b. p. =70.2 ℃)may be formed if water coexists with ethyl acetate and ethanol. Therefore, the unreacted ethanol and trace amount of water must be removed from ethyl acetate prior to the final distillation. Otherwise, the final yield and purity of ethyl acetate will be affected by the formation of azeotrope[③].

- **Questions**

1. What measures should be taken to drive the equilibrium in the esterification to the

---

[①] 有机相用饱和碳酸钠洗涤后，必须先用饱和食盐水洗涤，再用饱和氯化钙溶液洗涤。否则会产生絮状的碳酸钙沉淀，给后面的分离带来麻烦。

[②] 用饱和食盐水溶液代替水来洗有机相的目的有两个：一是防止乳化，二是降低乙酸乙酯因溶于水而造成的损失。

[③] 乙酸乙酯与水或乙醇会形成共沸物，若三者共存则生成三元共沸物。因此，最后一步蒸馏前，粗产物中未反应的醇和少量的水必须完全除去，以免形成共沸物，影响酯的收率和纯度。

right to improve the yield of esters?

2. What is the purpose of the first distillation after the completion of the esterification? What components does the distillate contain?

3. Why is it necessary to adjust the pH value of the mixture to 8 before washing the organic layer with saturated aqueous sodium chloride solution?

4. What is the purpose of each washing operation during the work-up procedure?

(1) Washing the organic layer with saturated aqueous sodium chloride solution.

(2) Washing the organic layer with 5.0 mL of saturated calcium chloride.

(3) Washing the organic layer with 5.0 mL of water.

5. Write down the stepwise mechanism of the acid-catalyzed esterification of acetic acid and ethanol; and use curved arrows to symbolize the flow of electrons.

Exp.17 Preparation of *n*-Dibutyl Phthalate

- **Objectives**

1. To study the principle and method of preparing esters by the reactions of acid anhydrides and alcohols.

2. To learn the technique of azeotropic removal of water using water segregator.

3. To practice the washing and drying of liquid compounds.

4. To master the technique of purifying liquid compounds by vacuum distillation.

- **Principle**

Although the acid-catalyzed Fischer esterification of a carboxylic acid and alcohol is the best known method for the preparation of an ester, the ester formation by Fischer esterification sometimes works inefficiently due to the existence of equilibrium in this reaction. In fact, esters are more generally prepared by the reactions of alcohols with acid chlorides or acid anhydrides in the laboratory[①]. For instance, *n*-dibutyl phthalate can be synthesized by the reaction of pathalic anhydride and *n*-butyl alcohol using concentrated sulfuric acid as acid catalyst. This conversion involves two steps:

$$\text{phthalic anhydride} + n\text{-}C_4H_9OH \xrightarrow{H_2SO_4} \text{o-}C_6H_4(COOC_4H_9)(COOH)$$

$$\text{o-}C_6H_4(COOC_2H_5)(COOH) + n\text{-}C_4H_9OH \underset{}{\overset{H_2SO_4}{\rightleftharpoons}} \text{o-}C_6H_4(COOC_4H_9)_2 + H_2O$$

The first step is a typical nucleophilic acyl substitution reaction, which proceeds quickly and completely, while the second step is actually a reversible Fischer esterification. In order to drive the equilibrium toward completion to forming ester, a water segregator is employed

---

[①] 在实验室，可以通过醇和酰氯或酸酐的反应来制备酯。

to remove the by-product water from the reaction mixture[66]. Concentrated sulfuric acid is used as a catalyst to promote the two-step conversions, which protonates the carbonyl-group oxygen atoms in the two substrates, and thereby renders them much more reactive towards the additions of weakly nucleophilic *n*-butyl alcohol.

- **Apparatus**

Apparatuses required include 100 mL three-necked flask, thermometer, thermometer adapter, water segregator, Allihn condenser, separatory funnel, graduate cylinder, beaker, 50 mL round-bottom flask and apparatus for vacuum distillation.

- **Setting Up**

Place 6 g of phthalic anhydride, 13 mL of *n*-butyl alcohol and 0.2 mL of concentrated sulfuric acid into a dried 100 mL three-necked flask, and mix the reactants thoroughly by gently swirling the flask. Add several zeolites into the flask and assemble the reaction apparatus as shown in Figure 3.21.

- **Experimental Procedures**

1. Add some water[67] into the water segregator.

2. Heat the reaction mixture slowly to reflux until the solid is dissolved.

3. Continue to heat the mixture to reflux gently at the temperature lower than 140 ℃ for about 1.5 h until the side neck of water separatory is filled with water[68].

4. Stop heating and cool down the reaction mixture to 70 ℃.

5. Transfer the liquid in the flask into a separatory funnel. Shake the funnel well and stand it for a moment until the content in the funnel is separated into two distinct layers.

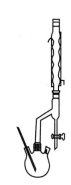

Figure 3.21 Apparatus for the preparation of *n*-dibutyl phthalate

6. Remove the aqueous layer and wash the organic layer sequentially with 10-15 mL 5 ％ of aqueous sodium carbonate solution and warm saturated aqueous sodium chloride solution(15 mL×2), with the pH of the organic layer kept at 7.

7. Pour the organic layer into a clean and pre-dried Erlenmeyer flask and dry the solution with anhydrous magnesium sulfate for 15-20 min.

8. Decant the dried solution into a pre-dried round-bottom flask carefully. Add several zeolites and assemble the flask for simple distillation to remove the *n*-butyl alcohol and other pre-distillate.

9. Assemble the apparatus for vacuum distillation, distill the final product, and then collect the pure *n*-dibutyl phthalate at 180-190 ℃ (10 mmHg) in a pre-weighed dry receiving

---

⑯ 反应分为两步：第一步是典型的酰基的亲核取代反应，此步反应是不可逆的。第二步实际上是可逆的费舍尔酯化反应。利用分水器来除去反应生成的副产物水，促进第二步反应向生成二丁酯的方向进行，提高产物的收率。

⑰ 分水器中加入水的量可由反应理论上生成水的量来决定。

⑱ 随着反应的进行，分水器中液面会慢慢升高。当分水器中全部被液体充满时，表明反应已经结束，此过程大约需要 1.5 h。

flask.

10. Weigh the final product and calculate the percentage yield of *n*-dibutyl phthalate.

• **Helpful Hint**

1. Phthalic anhydride is corrosive and irritating. Wear gloves and avoid skin contact while handling it. Perform this reaction in the hood to minimize exposure to it. Do not allow phthalic anhydride to come in contact with your skin. If it does, flush the affected area with copious amounts of water.

2. Concentrated sulfuric acid is corrosive and can cause severe burns. Be extremely careful while handling it. If any concentrated sulfuric acid comes in contact with your skin, immediately wash it off with copious amounts of cold water and then with dilute sodium bicarbonate solution.

3. The azeotropic boiling point of *n*-butyl alcohol and water is 92.7 ℃. The surface level of the liquid in the side neck of the water segregator will increase gradually with the evaporation and condensation of *n*-butyl alcohol and newly-formed water. The reaction should be completed when the surface of the liquid is at the top of side neck.

4. The reaction temperature will rise slowly as the reaction proceeds. Stop the reaction when the temperature reaches to 140 ℃. *n*-Dibutyl phthalate is easily decomposed under an acid condition when the temperature exceeds 180 ℃ [69].

$$\text{benzene-1,2-dicarboxylic acid dibutyl ester} \xrightarrow[180\ °C]{H^+} \text{phthalic anhydride} + \text{1-butene} + H_2O$$

5. When the organic layer is neutralized, keep the temperature within 70 ℃, and high concentration of base, such as sodium hydroxide cannot be used. Otherwise, the saponification of the *n*-dibutyl phthalate will occur easily [70].

6. Using saturated aqueous sodium chloride solution instead of water to wash the organic layer can reduce the loss of *n*-dibutyl phthalate due to its solubility in water. In addition, it can also prevent emulsification and contribute to layer separation [71].

• **Questions**

1. What side reactions may occur when heating *n*-butyl alcohol to high temperature in the presence of sulfuric acid?

2. What undesired reaction might occur if concentrated solution of sodium hydroxide is used to replace the aqueous sodium carbonate solution? Write down the corresponding reaction equation.

3. What is the purpose of each washing operation in the work-up procedure of this exper-

---

[69] 随着反应的进行，反应的温度不断升高。当温度上升至 140 ℃时即可停止反应。在酸性条件下，当温度超过 180 ℃时，邻苯二甲酸二丁酯就会发生分解反应生成邻苯二甲酸酐和 1-丁烯。

[70] 控制中和反应的温度不要超过 70 ℃，且不要使用浓度高的强碱。否则邻苯二甲酸二丁酯易发生皂化反应。

[71] 用饱和食盐水代替水来洗涤有机相，一方面是为了减少邻苯二甲酸二丁酯的损失，另一方面也是为了防止洗涤过程中乳化现象的发生。

iment?

(1) Washing the organic layer with 10-15 mL of 5 ％ sodium carbonate.

(2) Washing the organic layer with warm saturated aqueous sodium chloride solution(15 mL×2).

4. What will be the adverse effect of excess concentrated sulfuric acid on this reaction?

5. What will be the unexpected effect if the reaction mixture is heated vigorously at first?

3.7 Nitrogen-containing Compounds

Exp.18 Preparation of Acetanilide

- **Objectives**

1. To study the principle and method of preparing acetanilide by acylation reaction.

2. To learn the technique of fractional distillation.

3. To practice vacuum filtration, washing and recrystallization.

- **Principle**

The acylation of an amine is usually accomplished by a typical acyl nucleophilic substitution reaction of an amino group and acyl reagents, such as acyl halides or acid anhydrides. These acyl reagents are so reactive that the related acylation reaction is often accompanied by bis-acylation. Moreover, acyl halides or acid anhydrides are both expensive and sensitive to water, which limits their application in undergraduate laboratory. Although the acylation of an amine with a carboxylic acid proceeds slowly and requires long time of heating, this procedure is still of commercial interest. For instance, the acylation of aniline using acetic acid can produce acetanilide with relatively good yield. This reaction is reversible, and fractional distillation is applied to remove the by-product water once it is formed to drive this reaction toward completion[72].

$$\text{PhNH}_2 + \text{CH}_3\text{COOH} \underset{}{\overset{\text{Zn}}{\rightleftharpoons}} \text{PhNHCOCH}_3 + \text{H}_2\text{O}$$

The acetylation of the aryl amino group is reversible, which provides a method to protect a primary or a second amine[73]. The conversion of an amino group(—NH_2) into an amide group(—NHCOR) by acylation makes it possible to improve the selectivity of an electrophilic substitution reaction of aromatic amines, while the acyl group can be removed to regenerate the free amino group via acid hydrolysis upon the completion of the desired reaction[74].

---

[72] 本实验通过苯胺与乙酸的酰基化反应来制备乙酰苯胺。此反应是可逆的，且反应速率较慢。为了提高乙酰苯胺的收率，一般采用加入过量醋酸，同时利用分馏柱使反应中生成的水脱离反应体系的方法来促进平衡右移。

[73] 在有机合成中为了保护芳氨基，可先将芳胺乙酰化转化为乙酰芳胺，然后再进行其他反应，最后水解除去乙酰基而重新生成游离氨基。

[74] 通过酰化反应，可以大大提高芳胺亲电取代反应的选择性。芳氨基酰化后，由很强的第一类定位基变成了中等强度的第一类定位基，使芳环的亲电取代反应由多元取代变为了较易控制的一元取代。同时，由于乙酰基的空间位阻效应，往往选择性地生成对位取代产物。

- **Apparatus**

Apparatuses required include 100 mL Erlenmeyer flask, fractionating column, distillation head, thermometer, thermometer adapter, Büchner funnel, filter flask, graduate cylinder, and beaker.

- **Setting Up**

Add 5 mL of aniline, 7.4 mL of acetic acid and 0.1 g of zinc powder into a 100 mL Erlenmeyer flask. Mix the reactants thoroughly by gently swirling the Erlenmeyer flask and assemble the reaction apparatus as shown in Figure 3.22.

Figure 3.22 Apparatus for the preparation of acetanilide

- **Experimental Procedures**

1. Heat the reaction mixture to reflux for about 1h, with the temperature readout by the thermometer kept at about 105 ℃[75].

2. Upon the completion of the reaction, pour the hot solution into a beaker containing 100 mL of water while stirring[76].

3. Cool down the mixture to room temperature and filter the mixture under vacuum.

4. Collect the solid by vacuum filtration and wash it with 10 mL of water.

5. Purify crude acetylaniline by recrystallization using hot water as the solvent.

6. Collect the pure acetylaniline by vacuum filtration and dry it in the air.

7. Weigh the dried product and calculate the percentage yield of acetylaniline.

- **Helpful Hints**

1. Aniline is a highly toxic irritant, and can be absorbed by the skin. Wear gloves and avoid skin contact while handling it.

2. The color of aniline will become darkened after being stored for a long time due to the presence of impurities, which will affect the quality of acetylaniline. Therefore, aniline is preferably redistilled before use[77].

3. Acetic acid is corrosive and suffocate. Wear gloves and handle it in the hood. If any acid spills on your skin, wash it off with large amount of water.

4. Zinc powder is added to the reaction mixture to avoid the oxidation of aniline during the reaction progression[78].

5. The reaction temperature should be controlled carefully at about 105 ℃ to ensure the removal of the newly-formed water, but the starting material acetic acid remains in the reac-

---

[75] 反应过程中，要严格控制柱顶温度不要超过 105 ℃，以确保将生成的水及时脱离反应体系，同时防止反应物醋酸被蒸出。

[76] 反应结束后，要趁热将反应混合物倒入盛有冷水的烧杯中，待反应液冷却后，会立即析出固体，这些固体一旦粘在反应瓶壁上将难以倒出。

[77] 久置的苯胺因含有杂质而颜色变深，这会影响产物乙酰苯胺的质量。因而，苯胺在使用前最好重新蒸馏。

[78] 为了防止苯胺在反应过程中被氧化，可在反应混合物中加少许锌粉。

tion flask.

6. Given that the product will be precipitated immediately in the flask at lower temperature and thus cannot be easily poured out completely from the reaction flask, the reaction mixture should be poured into water when it is still hot.

7. The melting point of acetanilide is 114 ℃. Therefore, control the heating temperature and avoid "oiling out" during the recrystallization. If there are still some oil droplets not completely dissolved, a small amount of water can be added. Then, reheat the solution to boil until the oil droplets are dissolved[⑫].

- **Questions**

1. The acetylation of aniline by acetic acid is reversible. What measures should be taken in this experiment to drive the reaction to go forward?

2. Why must the temperature of distilling vapor be controlled to about to 105 ℃ in the acylation of aniline and acetic acid?

3. How many milliliters of water should be obtained at the end of the reaction according to the theoretical calculation? More liquid is always collected in the receiving flask than the theoretical amount. Explain the reason.

4. In addition to the sudden drop of temperature, are there any other observed phenomena that indicate the completion of the acylation reaction?

5. What are the requirements of a suitable solvent for recrystallization? What steps are involved in recrystallization?

6. The protection of the amino group has found many applications in organic synthesis. Try to summarize them and complete the following transformation.

Exp.19   Preparation of Methyl Orange

- **Objectives**

1. To study the principle and method of preparing azo compounds by diazo coupling reaction.

2. To practice filtration.

3. To understand how a pH indicator works.

- **Principle**

Methyl orange can be prepared by the diazo coupling reaction of diazotized sulfanilic acid and N,N'-dimethylaniline. The reaction equations are as follows:

---

⑫ 乙酰苯胺的熔点为114 ℃。粗产物乙酰苯胺在重结晶时，当水溶液加热至沸腾后，有时会发现未溶解的油珠漂浮在水面上。这些油珠是熔融状态下的含水乙酰苯胺，此时应补加水至油珠消失。

$$H_2N-\text{C}_6H_4-SO_3H + NaOH \longrightarrow H_2N-\text{C}_6H_4-SO_3Na + H_2O$$

$$H_2N-\text{C}_6H_4-SO_3Na + NaNO_2 + HCl \longrightarrow [HO_3S-\text{C}_6H_4-N{\equiv}N]^+Cl^- \xrightarrow{\text{C}_6H_5N(CH_3)_2 / HOAc}$$

$$[HO_3S-\text{C}_6H_4-N{=}N-\text{C}_6H_4-N^+(CH_3)_2H]\ OAc^- \xrightarrow{NaOH} NaO_3S-\text{C}_6H_4-N{=}N-\text{C}_6H_4-N(CH_3)_2$$

Diazotized sulfanilic acid, a kind of diazonium salt, is synthesized by adding nitrous acid (HNO_2) to sulfanilic acid, in which nitrous acid is generated in situ from sodium nitrite and hydrochloric acid. Although diazonium salts are safe when dissolved in an acidic solution at low temperature, they tend to decompose to release energy and nitrogen gas when heated or dried. Therefore, diazonium salts are always freshly prepared at lower temperature and must be used immediately without further isolation[30].

The diazonium ion is a weak electrophile, and then undergoes a coupling reaction with N,N'-dimethylaniline under weakly acidic condition to yield protonated cation of methyl orange in red, which is subsequently treated with sodium hydroxide to provide methyl orange[31].

Methyl orange is a widely used pH indicator. The aqueous solution of methyl orange with a concentration of 0.01% is usually used in titrations of acids due to its clear color changes and a sharp end point in the pH range of 3.1-4.4. In an acidic solution with a pH lower than 3.1, methyl orange appears as red protonated cation. Under basic conditions with a pH more than 4.4, methyl orange exists as the deprotonated anion in orange color[32].

- **Apparatus**

Apparatuses required include 50 mL beaker, 100 mL beaker, Büchner funnel, filter flask, graduate cylinder, stir rod, test tubes and Pasteur pipet.

- **Experimental Procedures**

1. Preparation of the dizaotized sulfanilic acid

(1) Place 1.1 g of sulfanilic acid and 5 mL of 5% sodium hydroxide solution into a 100 mL beaker[33].

(2) Stir the mixture with a stir rod and heat it in a warm-water bath until all the solids are dissolved.

(3) Allow the resulting solution to be cooled down to room temperature, and then add 0.4 g of sodium nitrite to the solution.

(4) Stir the mixture until all the solids disappear. Then, cool down the resulting solution in an ice-salt bath.

---

[30] 低温下，溶解在酸性溶液中的重氮盐是稳定的，但在加热或干燥时，重氮盐会分解并释放出氮气。因此，重氮盐的制备应在较低的温度下进行，且合成后必须立即使用，不需要进一步分离提纯。

[31] 偶合反应首先得到的是红色的酸式甲基橙，用碱处理后转化为橙色的甲基橙。

[32] 甲基橙的变色范围为 pH 3.1~4.4，pH<3.1 时变红，pH>4.4 时变黄，pH 3.1~4.4 时呈橙色。

[33] 对氨基苯磺酸是两性化合物，以酸性内盐的形式存在，不溶于盐酸，很难直接重氮化。因而，要采用倒重氮法，即先将对氨基苯磺酸溶在氢氧化钠溶液中，再加入需要量的亚硝酸钠，最后滴加盐酸。

(5) Add a mixture of 10 mL of cold water and 1.5 mL of concentrated hydrochloric acid dropwise to the cooled solution, with the reaction temperature kept below 5℃.

(6) Upon the completion of the addition, keep the suspension in the ice-water bath for 15 min to complete the reaction.

2. Preparation of methyl orange

(1) Add 0.6 g of N,N'-dimthylaniline and 0.5 mL of acetic acid into a 10 mL test tube and mix them thoroughly.

(2) Add the solution above dropwise to the cooled suspension of diazotized sulfanilic acid in the beaker while stirring.

(3) After the addition, keep the beaker in an ice-water bath and continue to stir the mixture vigorously to ensure the completion of the coupling reaction[㉞].

(4) Slowly add 12.5 mL of 5% sodium hydroxide to the mixture while stirring and check the pH of the solution constantly to make sure that the solution is basic. If not, add extra 5% sodium hydroxide until the color of the solution turns into orange and the crude methyl orange is precipitated.

(5) Heat the mixture to boil for 10 min to dissolve the crude methyl orange.

(6) Add 5 g of sodium chloride to the solution when all the solids disappear.

(7) Cool down the mixture in an ice-water bath to complete the precipitation after all the salts are dissolved.

(8) Collect the precipitated crystal by vacuum filtration, wash it with 20 mL of brine, and then dry the crystal in the air.

(9) Weigh the dried product and calculate the percentage yield of methyl orange.

(10) Methyl orange can be purified by recrystallization using hot water as solvent[㉟].

- **Helpful Hints**

1. Aromatic amines are harmful when ingested or inhaled. They can also be absorbed by the skin. Always wear gloves while handling these compounds. Work in the hood or in a well-ventilated area.

2. Hydrochloric acid and sulfuric acid are corrosive. If these acids come in contact with your skin, wash the affected area immediately with copious amounts of cool water.

3. Diazonium salts are explosive when dried. Do not keep the obtained diazonium solution for a long time and use it immediately in the next step.

4. The liquid waste containing diazonium salts cannot be poured directly into an aqueous waste tank. Treat the liquid with 1 g of potassium iodide and stir the resulting mixture for 20 min before disposing it into the aqueous waste container[㊱].

㉞ 偶合过程中，会有沉淀析出。因而反应过程中，反应混合物要充分、剧烈搅拌，以确保反应进行完全。

㉟ 甲基橙的重结晶操作要迅速，否则由于产物呈碱性，在较高温度下易发生变化，导致甲基橙的颜色加深。

㊱ 含有重氮盐的废液不能直接倒入盛放水溶液的废液缸中。处理重氮反应的废液时，应先向废液中加入适量的碘化钾并混合搅拌20 min，然后再倒入废液缸中。

5. Sulfanilic acid is a zwitterionic molecule and its acidity is slightly stronger than its basicity. Therefore, sulfanilic acid has poor solubility in water under an acid condition and its solubility increases significantly under the basic condition. However, it is necessary to carry out diazotization reaction under the acidic condition. To resolve this problem, sulfanilic acid is first dissolved in the solution of sodium hydroxide, the required amount of sodium nitrite is then added to the resulting solution, and then hydrochloric acid is finally added.

6. In the process of diazotization, the reaction temperature should be controlled below 5℃ to avoid the decomposition of the corresponding diazonium salts and the formation of the by-product of *p*-aminophenol[❷].

7. The presence of precipitates requires to vigorously and thoroughly stir the reaction mixture during the coupling reaction.

8. It might be helpful to add a bit of sodium hydroxide during the recrystallization of crude methyl orange. In order to avoid deterioration of the basic methyl orange at high temperature, the recrystallization of methyl orange should be performed quickly.

• **Questions**

1. Why must the diazotization reaction be carried out at low temperature? What side reactions may occur in the diazotization with sulfanilic acid? Write down the corresponding reaction equations.

2. In the diazotization step of this experiment, why must sulfanilic acid be dissolved in sodium hydroxide solution before being mixed with sodium nitrite?

3. At the end of the diazotization, starch-iodide paper is always used to determine whether additional sodium nitrite needs to be added to complete the reaction. Analyze the reason and write down the corresponding equation.

4. The coupling reaction between diazonium salts and arylamines should be carried out under weakly acidic conditions. Explain the reason.

5. Draw the structures of methyl orange under acidic and basic conditions respectively.

3.8 Heterocyclic Compounds

Exp.20   Preparation of 8-Hydroxyquinoline

• **Objectives**

1. To study the principle and method of preparing 8-hydroxyquinoline by Skraup reaction.

2. To practice steam distillation and recrystallization.

• **Principle**

8-Hydroxyquinoline is an important intermediate in organic synthesis and is extensively

---

❷ 重氮化过程中,要控制反应的温度。反应温度超过 5 ℃,重氮盐不仅易分解,且会发生水解生成苯酚,降低其收率。

used as staring material in the preparation of drugs and pesticides. It is also an excellent metal ion chelating agent used as extractant for metal ions.

8-Hydroxyquinoline can be obtained through the Skraup reaction[①] by co-heating 2-aminophenol with anhydrous glycerin in the presence of concentrated sulfuric acid and 2-nitrophenol.

$$\text{2-aminophenol} + \begin{array}{c} CH_2OH \\ CHOH \\ CH_2OH \end{array} \xrightarrow[\text{2-nitrophenol}]{H_2SO_4} \text{8-hydroxyquinoline}$$

The mechanism of this reaction is as follows: glycerol is firstly dehydrated by concentrated sulfuric acid to form propenal, which then undergoes conjugated nucleophilic addition with 2-aminophenol to produce the corresponding addition product. This addition product is then dehydrated by concentrated sulfuric acid and cyclized to generate 1,2-dihydroquinoline, which is subsequently oxidized by 2-nitrophenol to form the final product of 8-hydroxyquinoline[②].

$$\begin{array}{c} CH_2OH \\ CHOH \\ CH_2OH \end{array} \xrightarrow[-2H_2O]{H_2SO_4} \begin{array}{c} CHO \\ CH \\ \| \\ CH_2 \end{array}$$

$$\text{2-aminophenol} + \begin{array}{c} CHO \\ CH \\ \| \\ CH_2 \end{array} \longrightarrow \text{enamine intermediate} \xrightarrow[-H_2O]{H_2SO_4} \text{1,2-dihydroquinoline}$$

$$\text{1,2-dihydroquinoline} + \text{2-nitrophenol} \longrightarrow \text{8-hydroxyquinoline} + \text{2-aminophenol}$$

The mechanism above reveals that the structure of the oxidant nitro-aromatic hydrocarbons in Skraup reaction should be consistent with that of the substrate aromatic amines, otherwise by-products will be generated, thus causing troubles to the separation and purification of the final products[③].

• **Apparatus**

Apparatuses required include 150 mL three-necked flask, Allihn condenser, addition funnel, Erlenmeyer flask, graduate cylinders, apparatus for steam distillation and magnetic stirrer.

[①] Skraup 反应是合成喹啉及其衍生物最重要的方法之一。它是指由苯胺（或其他芳胺）与无水甘油、浓硫酸及弱氧化剂芳香族硝基化合物等共热合成喹啉的反应。

[②] Skraup 反应中，浓硫酸的作用是使甘油脱水生成丙烯醛，并使邻氨基苯酚与丙烯醛的加成物脱水成环。邻硝基苯酚为弱的氧化剂，能将成环产物 8-羟基-1,2-二氢喹啉氧化生成 8-羟基喹啉。

[③] Skraup 反应中所采用的氧化剂硝基芳烃的结构要和反应底物芳胺的结构保持一致，否则会形成副产物，给产物的分离和提纯带来困难。

- **Setting Up**

Place 1.8 g of 2-nitrophenol, 2.8 g of 2-aminophenol, 7.5 mL of anhydrous glycerin in sequence into a dry 150 mL three-necked flask. Mix the reactants thoroughly by swirling the flask and assemble the reaction apparatus as shown in Figure 3.23.

Figure 3.23 Apparatus for the preparation of 8-hydroxyquinoline

- **Experimental Procedures**

1. Cool down the reaction mixture in an ice-water bath.

2. Add 4.5 mL of concentrated sulfuric acid dropwise to the mixture from the addition funnel while stirring.

3. Upon the completion of the addition, slowly heat the mixture to reflux gently for 15 min.

4. Stop heating and remove the heating source, to make the reaction mixture smooth.

5. Heat the mixture again and reflux it gently for 1 h[①].

6. Stop heating and allow the resulting mixture to be cooled down. Then, add 15 mL of water to the mixture and assemble the reaction flask for steam distillation.

7. Distill the mixture until the distillate turns into colorless[②].

8. Cool down the reaction mixture, and slowly add 7 mL of sodium hydroxide solution to the flask. Then, carefully add 5 mL of saturated aqueous sodium carbonate to the mixture until the pH of the solution is 7-8[③].

9. Add 20 mL of water to the mixture, and re-assemble the reaction flask for steam distillation to distill the crude 8-hydroxyquinoline.

10. Cool down the obtained distillate completely and collect the solid by vacuum filtration.

11. Wash the solid with water and dry it in the air.

12. Purify crude 8-hydroxyquinoline by recrystallization using a mixture solvent of ethanol and water with the volume ratio of 4∶1.

13. Weigh the dried product and calculate the percentage yield of 8-hydroxyquinoline[④].

- **Helpful Hints**

1. 2-Nitrophenol and 2-aminophenol are sensitized and harmful if ingested or inhaled. Wear gloves while handling these compounds. Work in the hood or in a well-ventilated area. If these compounds come in contact with your skin, immediately wash them off with copious amounts of cold water.

2. Concentrated sulfuric acid is corrosive and can cause severe burns. Be extremely care-

---

[①] Skraup 反应是放热反应，反应的温度不能太高，以免反应过于剧烈，反应液冲出容器。

[②] 第一次水蒸气蒸馏的目的是蒸出没有反应完的邻硝基苯酚。

[③] 8-羟基喹啉既能与酸又能与碱反应生成盐，但一旦成盐后就不能经水蒸气蒸馏蒸出。所以，在第二次水蒸气蒸馏前，必须小心中和反应物，严格控制溶液的 pH=7~8。

[④] 产率以邻氨基苯酚计算，不考虑邻硝基苯酚被还原后参与反应的量。

ful while handling it. If any concentrated sulfuric acid comes in contact with your skin, immediately wash it off with copious amounts of cold water and then with dilute sodium bicarbonate solution.

3. All the glassware and chemicals used in this experiment must be adequately dried.

4. The amount of water in glycerol should be less than 0.5%. Otherwise, 8-hydroxyquinoline will be obtained in low yield[①].

5. Skraup reaction is exothermic, so the reactants should be added slowly to avoid explosion caused by vigorous reaction.

6. 8-Hydroxyquinoline can react with both acids and bases to form water-soluble salts, which are difficult to be distilled out by steam distillation. Therefore, the pH of the solution must be brought to 7-8 before the second steam distillation.

- **Questions**

1. Why must all the glassware and chemicals employed in Skraup reaction be completely dried?

2. Can nitrobenzene be used as the oxidant to replace o-nitrophenol for the preparation of 8-hydroxyquinoline through Skraup reaction? Explain your answer.

3. If 4-methylaniline replaces 2-aminophenol as the starting material, what product will be obtained through Skraup reaction? Write down the corresponding reaction equation.

4. There are two times of steam distillation during the work-up procedure of this experiment. Why must steam distillation be performed in acidic condition at the first time and under a neutral condition at the second time?

Exp.21 Extraction and Separation of Caffeine from Tea Leaves

- **Objectives**

1. To study the principle and operation of isolation natural components by solid-liquid extraction.

2. To learn how to use Soxhlet extractor.

3. To practice sublimation.

- **Principle**

Natural products are usually found in living organisms. Solid-liquid extraction is a valuable method for extracting a desired product from its natural source[②]. In this experiment, caffeine is isolated from tea leaves by solid-liquid extraction using ethanol as the solvent. Caffeine accounts for only about 5% of the mass of tea leaves. However, caffeine is extremely soluble in ethanol, which makes it relatively easy to be separated from the components insoluble in ethanol. Obviously, successive extraction can achieve better results, which can be achieved

---

[①] 反应所用甘油的含水量不能超过0.5%。如果甘油的含水量大，产物8-羟基喹啉的收率会降低。

[②] 固-液萃取法是从天然产物中提取所需物质的一种有效的方法。

fortunately with soxhlet extractor in the laboratory[20].

The natural source is usually composed of a variety of components and some of them share the similar properties, so that the solid-liquid extraction of natural materials often gets a mixture. In this case, additional operations are required to separate and purify further individual components. For example, tannin, which will be dissolved in ethanol, is always extracted with caffeine together from tea leaves. However, tannin is acidic and can be easily separated from caffeine by being treated with a base. Furthermore, caffeine can also be further purified by sublimation when heated at a high temperature.

Caffeine

- **Apparatus**

Apparatuses required include Soxhlet extractor, Allihn condenser, 250 mL round-bottom flask, 200 mL graduate cylinder, Liebig condenser, distillation head, thermometer with adapter, receiver adapter, Erlenmeyer flask, evaporating dish, asbestos pad and glass funnel.

- **Setting Up**

Add 60 mL of 95% ethanol and several zeolites into a round-bottom flask. Assemble the flask with soxhlet extractor and reflux apparatus as illustrated in Figure 3.24.

- **Experimental Procedures**

1. Place 10 g of tea leaves into a suitable filter cylinder, and then put the filter cylinder into a soxhlet extractor.

2. Heat the ethanol in the flask to reflux and continuously extract for 2 h until the color of the extracted liquid is very light[21].

3. Stop heating immediately when the extracted liquid just siphons back to the flask from the soxhlet extractor.

4. Assemble the flask with a simple distillation apparatus to distill most of the ethanol out until the residual liquid in the flask is about 5-8 mL.

Figure 3.24 Apparatus for isolation of caffeine from leaves using Soxhlet extractor

5. Transfer the residue onto an evaporating dish. Wash the flask with small amount of ethanol three times and add the washing liquid to the residue in the evaporating dish.

6. Add 2.5 g of CaO powder into the evaporating dish. Stir the mixture continuously until it is well mixed[22].

7. Put the resulting mixture in a boiling water bath until it is dried completely. Then, move the evaporating dish onto an asbestos pad.

---

[20] 在实验室，可以借助索氏提取器实现固-液萃取，索氏提取器利用溶剂的不断回流和虹吸作用，实现了多次、连续的萃取，从而达到最好的萃取效果。

[21] 控制加热回流的速度，一般2h内虹吸8~10次为宜。

[22] 生石灰起中和作用，可以除去茶叶中所含的单宁等酸性物质。

8. Cover a piece of large round filter paper with many small holes on the evaporating dish. Then, put an appropriately sized glass funnel upside down on the filter paper, and put a small amount of cotton into the neck of the funnel.

9. Sublime the crude caffeine with a small flame.

10. Stop heating once some brown oil substances appear on the wall of the glass funnel. Cool down and collect the caffeine crystal on the filter paper.

11. Repeat the above steps and re-sublime the residue caffeine on the evaporating dish twice with a slightly larger flame.

12. Put all the sublimated caffeine together. Weigh and calculate the percentage yield.

- **Helpful Hints**

1. During the extraction process, make sure that the filter cylinder is tightly close to the wall of soxhlet extractor. The height of the filter cylinder should be lower than that of the siphon arm with soxhlet extractor[⑩].

2. The mixture should be dried enough before being sublimated. If the mixture is not dry completely, some small drops of water will appear on the wall of the glass funnel at the beginning of sublimation. If so, stop heating, move the flame away, wipe the water drops off, and then continue the sublimation.

3. During the sublimation process, the temperature should be controlled carefully. Too high temperature may result in charred substances or make caffeine yellow, which will have a direct influence on the quality and yield of caffeine[⑪].

- **Questions**

1. What is the purpose of adding CaO to the mixture prior to the evaporation of this mixture to dryness?

2. What are the advantages of extracting desire component from a mixture with soxhlet extractor?

3. What criteria should be taken into account to determine whether to purify organic solid by recrystallization or sublimation?

4. Library project: search for other methods to extract caffeine from tea leaves.

---

⑩ 选择合适大小的滤纸筒，使其紧贴索氏提取器的壁，滤纸筒的高度不能超过索氏提取器的虹吸管。

⑪ 升华是提纯咖啡因的重要方法，操作的好坏将会直接影响到咖啡因的纯度和产量。在升华过程中要缓慢加热并严格控制温度。若温度过高，容易炭化或使产物变黄。

Chapter 4
Comprehensive Experiments

Exp.1 Preparation of Benzyl Alcohol and Benzoic Acid

- **Objectives**

1. To study the mechanism of the Cannizzaro reaction.
2. To practice the assembling of the reflux apparatus with a mechanical stirrer.
3. To practice extraction, washing, drying and simple distillation.
4. To practice the technique of purifying solid compounds by recrystallization.

- **Principle**

Cannizzaro reaction, also known as disproportionation reaction, refers to the redox reaction of an aldehyde with no α-H in the presence of a strong base, in which one molecule of aldehyde is oxidized to an acid while the other is reduced to an alcohol[102]. Sodium hydroxide and potassium hydroxide are the frequently used strong bases in Cannizzaro reaction.

Cannizzaro reaction takes place by the attack of hydroxide ion on the carbonyl group of an aldehyde to give a tetrahedral intermediate, followed by the transfer of a hydride ion to the carbonyl group of another aldehyde, in which both oxidation and reduction occur. Then an acid-base reaction occurs in the strongly basic medium to form the final products[103]. The accepted mechanism of the Cannizzaro reaction is as follows:

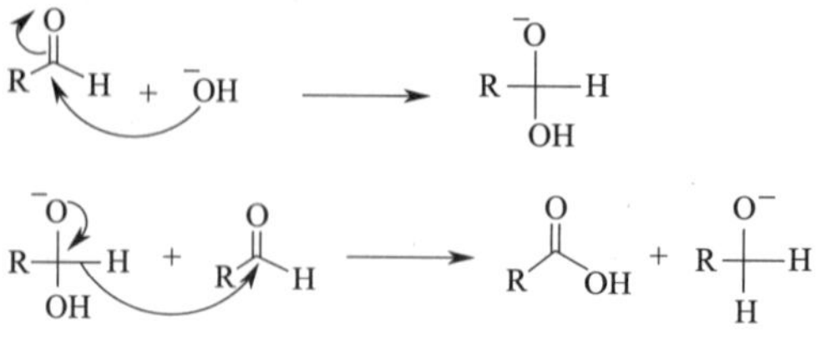

---

[102] Cannizzaro 反应（歧化反应）是指不含 α-H 的醛在浓碱的作用下，发生自身的氧化还原反应，即一分子醛被氧化成酸，另一分子醛被还原为醇。
[103] Cannizzaro 反应是连续两次的亲核加成反应。首先氢氧根负离子进攻一分子醛的羰基，生成正四面体加成产物。接着此加成产物提供氢负离子与第二分子醛的羰基进行加成，同时发生氧化和还原，最后经酸碱反应得到产物羧酸和醇。

In this experiment, benzoic acid and benzyl alcohol are prepared through the Cannizzaro reaction of benzaldehyde. The reaction scheme is demonstrated as follows:

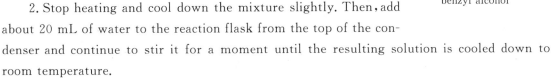

This reaction is a heterogeneous one due to the poor solubility of aldehydes in the concentrated aqueous solutions of sodium hydroxide. Accordingly, vigorous stirring is required to mix the reactants thoroughly. Meanwhile, high reaction temperature is also involved for the completion of the reaction.

- **Apparatus**

Apparatuses required include 250 mL three-necked flask, Allihn condenser, Büchner funnel, filter flask, separatory funnel, 100 mL round-bottom flask, distillation head, Liebig condenser, thermometer, thermometer adapter, receiver adapter, Erlenmeyer flask, graduate cylinder and apparatus for mechanical stirrer.

- **Setting Up**

Add 8 g of sodium hydroxide and 30 mL of water into a 250 mL three-necked flask and stir the mixture until the solid is dissolved. Cool down the solution to room temperature, and then add 10 mL of benzaldehyde into the reaction flask. Assemble the apparatus for reaction as shown in Figure 4.1.

Figure 4.1 Apparatus for the preparation of benzoic acid and benzyl alcohol

- **Experimental Procedures**

1. Heat the reaction mixture in the flask to reflux④ for about 40 min while stirring vigorously⑤ until the solution becomes clear.

2. Stop heating and cool down the mixture slightly. Then, add about 20 mL of water to the reaction flask from the top of the condenser and continue to stir it for a moment until the resulting solution is cooled down to room temperature.

3. Transfer the aqueous solution to a separatory funnel. Extract the solution with 10 mL of diethyl ether two times and combine the ether layers in an Erlenmeyer flask while collecting the aqueous layer in a beaker.

---

④ 苯甲醛的 Cannizzaro 反应要在加热回流的条件下进行。反应过程中要控制适当的反应温度。如反应温度太低则反应速率慢，转化率低。但反应的温度也不能太高，以免苯甲醛被氧化。

⑤ 苯甲醛的 Cannizzaro 反应是在两相间进行的，必须搅拌，使反应物接触更充分，加快反应的进行。

4. Wash the ether layer sequentially with 5 mL of saturated aqueous sodium bisulfite[06], 10 mL of 10% aqueous sodium carbonate and 10 mL of water.

5. Transfer the resulting ether layer into a clean Erlenmeyer flask and dry it with several amount of anhydrous magnesium sulfate. Occasionally swirl the Erlenmeyer flask until the solution becomes clear.

6. Decant the ether solution into a dry round-bottom flask and assemble the flask for simple distillation to distill diethyl ether[07].

7. Continue to heat the solution in the flask. Replace the Liebig condenser with an air condenser when the temperature reaches 140 ℃[08], and collect the fraction boiling between 198-204 ℃(760 mmHg) in a pre-weighed dry receiving flask.

8. Weigh the product and calculate the percentage yield of benzyl alcohol.

9. Allow the saved aqueous layer to be cooled down in an ice-water bath and acidify the aqueous solution by slowly adding a mixture of 30 mL of concentrated hydrochloric acid and 30 mL of water while stirring vigorously.

10. Collect the solid by vacuum filtration. Wash the filter cake with 5-10 mL of water and dry it in the air.

11. Purify the crude benzoic acid by recrystallization with hot water.

12. Weight the product and calculate the percentage yield of benzoic acid.

- **Helpful Hints**

1. The concentrated solution of sodium hydroxide is highly corrosive and caustic. Wear latex gloves while preparing and transferring it. Do not allow sodium hydroxide solution to contact with your skin. If it does, flood the affected area immediately with water and then thoroughly rinse it with 1% acetic acid.

2. Benzaldehyde is toxic and irritating. Wear gloves and avoid skin contact while handling it. Benzaldehyde is easily oxidized to benzoic acid and should be re-distilled before use.

3. When the reaction solution changes from cloudy to clear, it indicates the completion of the reaction[09].

4. Washing the ether layer with saturated solution of sodium bisulfite helps to remove the unreacted benzaldehyde.

5. In simple distillation, an air condenser should be applied to collect the fractions at a boiling point above 140 ℃.

6. Concentrated hydrochloric acid should be added slowly to the aqueous solution while

---

[06] 用饱和亚硫酸氢钠溶液洗涤醚相的目的是除去没有反应完的苯甲醛。

[07] 乙醚沸点低，容易着火。为保证安全，使用乙醚时近处不能有明火。在蒸馏乙醚时可在接引管上连接一长橡皮管，通入水槽的下水道，接收瓶要用冷水冷却。

[08] 简单蒸馏时，若被蒸馏的物质沸点高于140℃，蒸气的温度高，直形冷凝管的内管与外管结合处易发生爆裂，此时应改用空气冷凝管。

[09] 苯甲醛在水溶液中溶解度小，随着反应的进行，一部分苯甲醛被氧化生成水溶性的苯甲酸的钠盐，另一部分苯甲醛被还原生成在水中溶解度相对比较大的苯甲醇，反应液会逐渐由浑浊变澄清。因而，可以用反应液由浑浊变澄清来判定反应的终点。

stirring to avoid the risk of vigorous reaction.

- **Questions**

1. Why does the Cannizzaro reaction occur much more slowly in dilute than in concentrated sodium hydroxide solution?

2. What is the purpose of each extraction or washing operation in the work-up procedures of this experiment? Write down the corresponding equations for the chemical change of (2) and (3).

(1) Extracting the aqueous layer with ether (10 mL×3).

(2) Washing the ether layer with saturated sodium bisulfite solution.

(3) Washing the ether layer with 10% Na_2CO_3.

(4) Washing the ether layer with water.

3. No deuterium is found on the benzylic carbon atom in the benzyl alcohol formed when the Cannizzaro reaction is performed with benzaldehyde in D_2O solution. Explain the reason.

4. Write down an equation for the crossed Cannizzaro reaction of benzaldehyde and formaldehyde in the presence of concentrated sodium hydroxide solution.

5. Discuss the differences observed in the IR and ^1H NMR spectra of benzaldehyde, benzyl alcohol, and benzoic acid that are consistent with the formation of the two products from benzaldehyde by the Cannizzaro reaction.

Exp.2 Preparation of Benzil and Thin-layer Chromatography

- **Objectives**

1. To study the method of preparing benzil via the oxidation of benzoin.

2. To practice refluxing and recrystallization.

3. To learn to produce thin-layer chromatography plates.

4. To learn to monitor the progress of a reaction by thin-layer chromatography analysis.

- **Principle**

1,2-Diphenylethane-1,2-dione, also known as benzil[①], is one of the most common α-diketones that can be utilized as an important intermediate in pharmaceuticals and organic synthesis. Benzil can be easily prepared by oxidation of benzoin[②] with a variety of oxidizing agents, such as nitric acid, copper sulfate in pyridine, $Cu(OAc)_2/NH_4NO_3$, etc. as well as some metallorganic oxidants. Although these oxidants work efficiently for the conversion of benzoin to benzil, some of the methods may be accompanied by vigorous reactions, tedious work-up procedures, toxic regents or low conversion yields.

In this experiment, ferric trichloride is used as the oxidizing agent for the conversion of benzoin to benzil due to its operability and consistent results. The reaction scheme is demon-

---

[①] 苯偶酰。

[②] 苯偶姻，安息香。

strated below. Acetic acid is hereby used not only as a solvent, but also to provide an acidic medium for this reaction to inhibit the hydrolysis of ferric chloride⑪

$$\text{Ph-CO-CH(OH)-Ph} + FeCl_3 \xrightarrow[\Delta]{CH_3COOH} \text{Ph-CO-CO-Ph} + FeCl_2 + HCl$$

Thin-layer chromatography (TLC)⑫ is a simple, inexpensive and efficient method to characterize organic compounds for purity and identities, and has been extensively adopted in organic laboratory. In this experiment, TLC analysis is used to monitor the progress of the oxidation⑬ of benzoin and help to accurately determine the end point of this reaction.

- **Apparatus**

Apparatuses required include 100 mL three-necked flask, Allihn condenser, Büchner funnel, filter flask and Erlenmeyer flask.

- **Setting Up**

Place 10 mL of acetic acid, 10 mL of water and 5.5 g of ferric trichloride into a 100 mL three-necked round-bottom flask. Mix the solution thoroughly and add several zeolites into the flask. Assemble the reaction apparatus as shown in Figure 4.2.

- **Experimental Procedures**

1. Preparation of benzil

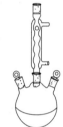

Figure 4.2 Apparatus for the preparation of benzil

(1) Slowly heat the reaction mixture in the flask until the mixture begins to boil.

(2) Stop heating and add 2.2 g of α-hydroxyketone (benzoin) to the flask when the mixture is no longer boiling.

(3) Heat the mixture in the flask to boil again and record the time.

(4) Monitor the progress of the reaction by TLC analysis. Spot the TLC plate on the base line with a small amount of the reaction mixture after the reaction has heated to reflux for 15 min. Also, spot the same plate on the base line with the dilute ethanol solution of starting material benzoin. Then, develop this chromatogram in the developing chamber using dichloromethane as the developing solvent.

(5) Repeat the refluxing/spotting process when the reaction is heated to reflux for 45 min.

(6) Add 50 mL of water to the reaction flask from the top of the condenser after the reaction is completed, and then heat the resulting mixture to boil for a while.

(7) Allow the mixture to cool down to room temperature until the yellow solid is completely precipitated out.

---

⑪ 醋酸不仅可以作为反应的溶剂，也为该反应提供了酸性的介质，以抑制三氯化铁的水解。

⑫ 薄层色谱。

⑬ 追踪反应进程。

(8) Collect the solid by vacuum filtration and wash it with cold water.

(9) Purify the crude benzil by recrystallization with 80% ethanol.

(10) Weigh the dried product and calculate the percentage yield of benzil.

(11) Measure the melting point of benzil to characterize the product.

2. The general produce for producing TLC plates

(1) Weight 5.0 g of GF254 silica gel⑮, and mix it thoroughly with 10 mL of 0.5% CMC solution⑯ in a small beaker.

(2) Spread the prepared silica gel mud on two clean and dried plates with the layer of 1-2mm thick. The thickness of the layer should be as uniform as possible.

(3) Dry the two TLC plates at room temperature for a moment, and then activate them in an oven for an hour at 110 ℃ to remove any adsorbed water.

(4) Cool down the two TLC plates to room temperature before use.

- **Helpful Hints**

1. Acetic acid is corrosive and suffocative, so handle it in the hood and wear gloves. If any acid spills on your skin, wash it off with large amount of water.

2. Ferric trichloride is corrosive, so wear gloves while handling it. In addition, ferric chloride can deliquesce easily, weigh and transfer it as fast as possible⑰.

3. Recrystallization of crude benzil cannot be performed in a beaker, and reflux apparatus should be used to dissolve the crude benzil in 80% ethanol to form a saturated solution⑱.

4. Tips for thin-layer chromatography analysis:

(1) All the glassware for producing TLC plates should be dry and clean.

(2) Spot the TLC plate with a small amount of dilute(1%-2%) instead of too much sample solution, which will lead to large tailing spots and poor separation⑲.

(3) The sample applied on the TLC plate must be higher than the surface of the developing solvent in the developing chamber to avoid the sample from being dissolved in the developing solvent⑳.

(4) The chamber must be saturated with solvent vapor during the whole progress of thin lay chromatography analysis㉑.

(5) Mark the solvent front on the adsorbent immediately when the plate is removed from the chamber.

(6) Calculate the R_f value of the spots on the TLC plates.

---

⑮ 既含有煅石膏黏合剂又含有荧光剂的硅胶。

⑯ 羧甲基纤维素钠是一种黏合剂。

⑰ 三氯化铁很容易潮解，量取时动作要快。

⑱ 二苯基乙二酮可以在80%的乙醇中重结晶提纯。重结晶的操作不能在烧杯中进行，加热制备热的饱和溶液时应加装回流装置。

⑲ 点样时，样品的浓度不要过大，否则层析时会出现拖尾现象，影响分离的效果。

⑳ 层析缸里不要加太多的展开剂，展开剂的液面要低于薄层色谱板上样品点的高度，以免样品溶解在展开剂中。

㉑ 薄层色谱的展开要在密闭的容器中进行，展开的过程要盖上层析缸的盖子，保证层析缸始终被溶剂蒸气饱和。

• **Questions**

1. What is the purpose of adding acetic acid to the reaction mixture in this experiment?
2. Discuss the differences observed in the IR and ^{1}H NMR spectra of benzil and benzoin.
3. List the main steps for recrystallization, and briefly explain the purpose of each step.
4. What is the principle of TLC analysis? What is the TLC analysis used for in organic synthesis?
5. How many indirect visualization techniques can be used in TLC analysis for colorless compounds? Briefly describe the principles of these visualization techniques.

Exp.3 Preparation of Acetylferrocene and Column Chromatography

• **Objectives**

1. To study the principle and method of introducing an acyl group into an aromatic rings by Friedel-Crafts acylation reaction.
2. To learn the technique of separating different components of a mixture by column chromatography.
3. To practice assembling the reflux apparatus with a mechanical stirrer.
4. To learn to remove solvents with a low boiling point by rotary evaporator.
5. To practice washing and drying solid compounds.

• **Principle**

Ferrocene is a well-known sandwich coordinate complex, in which the two cyclopentadienyl rings coordinate to a ferrous ion. The two cyclopentadienyl rings are aromatic and highly electron rich, and accordingly undergo some electrophilic substituted reaction with higher reactivity than benzene[⑫]. For example, ferrocene can undergo Friedel-Crafts acylation reaction with acetic anhydride under a milder condition, with phosphoric acid taken as the catalyst, which mainly produces monosubstituted acetyl ferrocene and a small amount of by-product 1,1′-diacetylferrocene. The reaction is shown as follows.

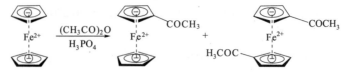

Column chromatography is used efficiently to separate and purify organic compounds on a larger scale from milligrams to hundreds of grams. In column chromatography, the stationary phase is solid adsorbents (silica gel or aluminum oxide) packed into a column, while an eluting solvent serves as the mobile phase. When the eluting solvent is added from the top of the column, the sample initially adsorbed on the adsorbent at the top of column will begin to move down the column along with the eluting solvent. The interactions of the individual com-

---

⑫ 二茂铁具有夹心的结构，由两个环戊二烯负离子环和一个二价铁离子键合而成。两个环戊二烯环具有芳香性，且比苯更容易发生亲电取代反应。

ponents of the sample with the stationary phase and the mobile phase determine the rate at which the different components elute from the column. In general, the more weakly the component is absorbed, the more rapidly it will be eluted. Thus, with a polar adsorbent, the less polar component will travel faster down the column than the more polar components[22].

After the Friedel-Crafts acylation reaction of ferrocene, a mixture of acetyl ferrocene, 1,1′-diacetylferrocene and ferrocene is always obtained. Column chromatography is applied to separate the desired product acetyl ferrocene from this mixture based on the different adsorption capacities of individual components with silica gel. All the components to be separated in this experiment are highly colored, which makes the column chromatography separation visually observed.

- **Apparatus**

Apparatuses required include 100 mL three-necked flask, Claisen adapter, addition funnel, thermometer, Allihn condenser, Erlenmeyer flask, thermometer, thermometer adapter, beaker, graduate cylinder, Büchner funnel, filter flask, rotary evaporation, apparatus for mechanical stirrer and column chromatography separation.

- **Setting Up**

Place 1.5 g of ferrocene and 10 mL of acetic anhydride into a 100 mL three-necked flask. Assemble the reaction apparatus as shown in Figure 4.3.

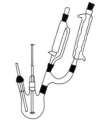

Figure 4.3 Apparatus for the preparation of acetyl ferrocene

- **Experimental Procedures**

1. Preparation of acetyl ferrocene

(1) Put the reaction flask into an ice-water bath and turn on the stirrer.

(2) Place 2 mL of 85% phosphoric acid in the addition funnel. Add phosphoric acid dropwise to the flask, with the temperature kept below 20 ℃[24].

(3) Upon the completion of the addition, stir the mixture at room temperature for 5 min, heat the mixture to 55-60 ℃, and then continuously stir it for another 15 min[25].

(4) Stop heating and stirring, pour the mixture into a beaker containing 40 g of crushed ice while it is still hot.

(5) Rinse the reaction flask with a little amount of cold water and combine the washing solution to the beaker[26].

(6) Add about 10 g of sodium carbonate in batches to the mixture while stirring vigor-

---

[22] 样品中的各组分与固定相和流动相之间的相互作用决定了不同组分从柱子上被洗脱下来的速度。一般来说，吸附能力较弱的组分，先随洗脱剂流出。因此，当使用极性吸附剂时，极性较弱的组分比极性强的组分优先被洗脱。

[24] 将磷酸加入二茂铁的乙酸酐溶液的过程中会放出大量的热，因此，磷酸一定要在冷却且充分搅拌下缓慢滴加，否则易产生深棕色黏稠状氧化聚合物。

[25] 一定要严格控制反应的温度在55～60℃，反应结束后，反应物呈暗红色。若温度高于85℃，反应物会发黑、黏稠，甚至炭化。

[26] 乙酰二茂铁在水中有一定的溶解度，洗涤时要用冰水，洗涤次数也切忌过多。

ously until the solution is neutral(pH=7)[17].

(7) Put the beaker in an ice-bath and allow the resulting solution to be cooled down to precipitate all the solid.

(8) Collect the yellow solid by vacuum filtration and wash it twice with cold water.

(9) Dry the crude acetyl ferrocene under infrared light.

2. Column chromatography separation

(1) Packing a column with the wet-packing method[18]

① Clamp a clean and dry chromatography column in a vertical position onto an iron stand.

② Close the stopcock of the column and add petroleum ether to one-third of the height of the column.

③ Put 30 g of silica gel slowly into a beaker with an excess of petroleum ether (about 1.5 times of the volume of silica gel). Then, mix the solvent and silica gel thoroughly and rigorously to remove the air present in the silica gel slurry.

④ Add the pre-mixed slurry of silica gel into the column in batches through a funnel. The slurry should be swirled thoroughly before adding each portion to the column.

⑤ Rinse and flush the slurry left in the beaker into the column using a small amount of petroleum ether. Rinse down the silica gel sticking to the wall of the column.

⑥ Tap the side of the column constantly as the slurry flows down the column. Then, add 5 mm of quartz sand carefully to the top of the silica gel after all the silica gel is settled.

⑦ Apply pressure to pack the column firmly until only a small amount of solvent is left above the quartz sand. More fresh petroleum ether may be needed to keep the solvent level always falling above the level of the adsorbent at any time.

⑧ Close the stopcock and get prepared for loading the sample to the column.

(2) Applying the sample to the column

① Weigh 0.1 g of the crude product into a dry 25 mL of beaker and add ethyl acetate until the solid is dissolved[19].

② Add 1.0 g of silicagel to the sample solution and mix them thoroughly, and then dry the orange slurry under infrared light.

③ Add the dried powdered solid to the top of the column, and rinse the sample stuck to the wall of the column with a small amount of petroleum ether.

④ Tap the side of the column to make the top of sample horizontal, and then add 3 mm of quartz sand to the top of the sample to prevent the adsorbent and sample from being disturbed.

⑤ Open the stopcock and allow the upper level of the solvent just above the top of sand.

---

[17] 用碳酸钠中和粗产物时，会有大量泡沫生成。为了防止产物随泡沫溢出，碳酸钠要边搅拌边慢慢分批加入。

[18] 装柱要紧密、无断层、无缝隙、无气泡，装柱的过程要始终保持有溶剂覆盖吸附剂。

[19] 乙酸乙酯的用量以样品恰好溶解为宜。

Then close the stopcock to stop the flow⑩.

(3) Elution of the sample

① Add a mixture of 5 : 1 petroleum ether/ethyl acetate to the column slowly and carefully, so that the upper layer of the column will not be disturbed.

② Open the stopcock of the column and collect the eluent with an Erlenmeyer flask.

③ Continuously add the eluting solvent to make sure that the column never dries out and maintain an optimum elution rate of 1~2 mL per minute⑪.

(4) Collection of fractions

① Three separate bands appear in the silica gel as the elution solvent moves down the column, among which, the upper one is in dark brown, the middle one is in orange and the lower one is in yellow.

② When the yellow band is at the bottom of the column, change the collection flask and collect the colored effluent with a dry Erlenmeyer flask⑫.

③ When the yellow band has been eluted from the column, collect the colorless effluent between the yellow band and the orange band with another Erlenmeyer flask.

④ When the orange band is at the bottom of the column, collect the effluent with a pre-weighted dry 100 mL round-bottom flask.

(5) Collection of the product

① Recover the pure acetyl ferrocene by removing the eluting solvent using a rotary evaporator.

② Dry the product under infrared light. Weigh and calculate the percentage yield of acetyl ferrocene.

(6) Cleaning of columns

① Drain out the remaining solvent in the column. Invert the column and push out the silica gel to a beaker.

② Discard the dried silica gel to the solid waste container.

- **Helpful Hints**

1. Phosphoric acid is corrosive. Wear gloves while handling it.

2. Acetic anhydride is corrosive and irritating. Wear gloves and avoid skin contact while handling it. Perform this reaction in the hood to minimize exposure to it. Do not allow acetic anhydride to come in contact with your skin. If it does, flush the affected area with copious amounts of water.

3. Phosphoric acid must be added dropwise to the cooled mixture of ferrocene and acetic anhydride. Otherwise, some side reactions may take place and bring about sticky by-products in dark brown.

4. When the crude product is neutralized with sodium carbonate, a large amount of foam

---

⑩ 加入洗脱剂前,应确保柱内溶剂的上层刚好在石英砂的顶部,不要留过多溶剂。
⑪ 样品的洗脱过程要始终保持有溶剂覆盖吸附剂。
⑫ 为减少溶剂的用量,洗脱时黄色带流出前的大部分洗脱液可作为纯的石油醚回收使用。

will be produced. In order to prevent the product from overflowing, sodium carbonate should be added in batches carefully while stirring.

5. Petroleum ether is highly volatile and flammable. Make sure that there are no flames in the vicinity while handling it.

6. Air bubble, crack or channel and irregular surface in the column may lead to poor separation. Never let the column dry out the solvent during the column chromatography separation progress.

7. Maintain the flowing rate of the eluting solvent at 1-2 mL/min. A too slow flowing rate will cause the diffusion of the component bands and result in poor separation[⑫].

• **Questions**

1. Try to explain why the acylation reaction never occurs more than once on an aromatic ring?

2. Ferrocene is more susceptible to electrophilic substitution reaction than benzene. Why is it difficult to get a desired product when ferrocene is nitrated with a mixture of concentrated nitric acid and sulfuric acid.

3. What are the advantages and disadvantages of column chromatography separation? What makes the acylation of ferrocene particularly attractive for column chromatography separation?

4. In the column chromatography separation, the order of changing eluting solvents must be from less polar solvent to a more polar one. Explain the reason.

5. Discuss the differences observed in the IR and ^{1}H NMR spectrum of acetyl ferrocene and ferrocene that are consistent with the conversion in this experiment.

Exp.4 Preparation of Ethyl Acetoacetate by the Claisen Condensation Reaction

• **Objectives**

1. To understand the mechanism of Claisen condensation reaction.
2. To study the method of preparing ethyl acetoacetate by Claisen condensation reaction.
3. To learn to carry out a reaction under anhydrous condition.
4. To practice the extraction, washing and drying of the liquid compounds.
5. To practice simple distillation and vacuum distillation.

• **Principle**

In the presence of a basic catalyst, an ester containing α-H can undergo a reversible carbonyl condensation reaction with another molecule of ester to form a β-keto ester. This reaction between two ester molecules is known as the Claisen condensation reaction[⑬], which is useful and synthetically important in organic synthesis. For instance, ethyl acetoacetate can

---

⑫ 控制洗脱剂流出的速度为 1～2d/s。洗脱速度过慢，会引起色带的扩散，导致分离效果不佳。

⑬ 含有 α-H 的酯在碱性催化剂存在的条件下，能与另外一分子的酯发生缩合反应生成 β-酮酸酯。发生在两分子酯之间的此类反应被称作克莱森缩合反应。

be prepared via the Claisen condensation reaction of two equivalents of ethyl acetate under a basic condition. The corresponding reaction is shown as follows.

$$2CH_3COC_2H_5 \xrightarrow[\text{2. }CH_3COOH]{\text{1. }NaOC_2H_5/C_2H_5OH} H_3C-\underset{O}{\overset{O}{C}}-CH_2-\underset{}{\overset{O}{C}}-OC_2H_5 + C_2H_5OH$$

The mechanism of the Claisen condensation involves an ester enolate ion formed by the base abstracting an acidic α-H atom from an ester molecule, which is added in a nucleophilic addition reaction to a second ester molecule, rendering a tetrahedral alkoxide intermediate. This intermediate expels alkoxide ions to yield a β-keto ester. Due to its high acidity, the produced β-keto ester will be easily deprotonated by the basic alkoxide ion. At last, a final β-keto ester is obtained after the protonation of the enolate ion obtained in the previous step by the addition of aqueous acid. Each step of the Claisen condensation reaction is reversible, and only the deprotonation step drives the equilibrium completely to the product side and leads to high yields in Claisen condensations[⑮].

(1) $CH_3COC_2H_5 \underset{}{\overset{-OC_2H_5}{\rightleftharpoons}} {}^-CH_2-\overset{O}{C}-OEt + C_2H_5OH$

(2) $H_3C-\underset{OC_2H_5}{\overset{O}{C}} + {}^-CH_2-\overset{O}{C}-OEt \rightleftharpoons H_3C-\underset{OC_2H_5}{\overset{O^-}{\underset{}{C}}}-CH_2-\overset{O}{C}-OC_2H_5$

(3) $H_3C-\underset{OC_2H_5}{\overset{O^-}{\underset{}{C}}}-CH_2-\overset{O}{C}-OC_2H_5 \rightleftharpoons H_3C-\overset{O}{C}-CH_2-\overset{O}{C}-OC_2H_5 + C_2H_5O^-$

(4) $H_3C-\overset{O}{C}-CH_2-\overset{O}{C}-OC_2H_5 + C_2H_5O^- \rightleftharpoons H_3C-\overset{O}{C}-\overset{-}{C}H-\overset{O}{C}-OC_2H_5 + C_2H_5OH$
$$\downarrow H_3O^+$$
$$H_3C-\overset{O}{C}-CH_2-\overset{O}{C}-OC_2H_5$$

As the analytical grade ethyl acetate always contains a small amount of ethanol, ethyl acetate and sodium metal are used as raw materials to prepare ethyl acetoacetate in this experiment. Sodium metal is easy to react with water, thus giving off hydrogen gas and a lot of heat, which may also lead to combustion and explosion. In this case, the apparatus used for this reaction must be dried completely and the reagents must be absolutely anhydrous.

● **Apparatus**

Apparatuses required include 100 mL round-bottom flask, Allihn condenser, drying tube containing anhydrous calcium chloride, separatory funnel, 50 mL round-bottom flask, distillation head, Liebig condenser, thermometer, thermometer adapter, receiver adapter and apparatus for vacuum distillation.

---

⑮ Claisen 缩合反应中，每一步反应都是可逆的，只有去质子化的一步即反应机理中的第（4）步将平衡推向生成产物的一侧，从而使 Claisen 缩合反应能获得较高的收率。

• **Setting Up**

Place 27.5 mL of ethyl acetate and 2.5 g of sodium metal thin flakes in a dry 100 mL round-bottom flask. Assemble the flask with a reflux apparatus with a drying tube containing anhydrous calcium chloride attaching to the top of the condenser. The reaction apparatus is shown in Figure 4.4.

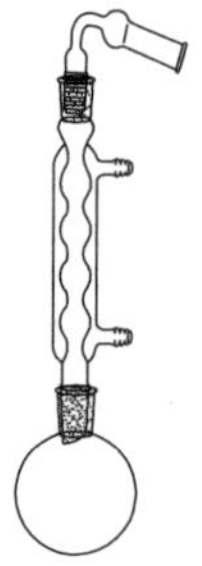

• **Experimental Procedures**

1. Heat the reaction mixture to reflux gently for about 2 h until the sodium metal disappears[⑮]. The solution in the flask should be clear red.

Figure 4.4 Apparatus for the preparation of ethyl acetoacetate

2. Stop heating and allow the mixture to cool.

3. Add 50% acetic acid slowly to the mixture while stirring until the solution is slightly acidic(pH=5-6)[⑯].

4. Transfer the reaction mixture to a separatory funnel and add an equal volume of saturated aqueous sodium chloride solution. Shake the funnel thoroughly, and then let it stand for a moment until the content in the funnel is separated into two distinct layers.

5. Save the organic layer and extract the aqueous layer with 5 mL of ethyl acetate.

6. Combine the ethyl acetate extracts with the saved organic layer. Wash the resulting organic layer with 5% sodium carbonate solution until the solution becomes neutral(pH=7).

7. Transfer the organic liquid into a clean and dry Erlenmeyer flask and dry the solution with anhydrous magnesium sulfate. Occasionally swirl the Erlenmeyer flask until the solution becomes clear.

8. Decant the dried solution into a dry round-bottom flask and assemble the flask for simple distillation to remove the ethyl acetate.

9. Assemble the apparatus for vacuum distillation and collect the pure ethyl acetoacetate in a pre-weighed dry receiving flask[⑰].

10. Weigh the product and calculate the percentage yield of ethyl acetoacetate. Estimations of the boiling points of ethyl phenyl ether at different pressures are shown in Table 4.1.

Table 4.1 Estimations of the boiling points of ethyl phenyl ether at different pressures[⑱]

Pressure/mmHg	5	10	20	40	60	100	200	400	760
Boiling point/℃	43	56	70	87	96	108	128	150	172

---

⑮ 反应如果太过剧烈，可停止加热并将反应瓶放入冷水浴中冷却，使反应平稳。

⑯ 醋酸会和反应混合物中没有反应完的金属钠剧烈作用，因而滴加醋酸时要缓慢，且应边加边搅拌，加至溶液呈微酸性即可。如果此时反应液中仍有少量固体未溶解，可以随反应液一起转移至分液漏斗中，然后加入饱和食盐水混合振摇，固体会自行溶解。醋酸不可多加，醋酸滴加过量，溶液的酸度过高，产物在水中的溶解度会增大，最终将导致乙酰乙酸乙酯收率的降低。

⑰ 通过减压蒸馏纯化乙酰乙酸乙酯。

⑱ 乙酰乙酸乙酯在受热温度超过95 ℃时就分解，所以应通过减压蒸馏来提纯。减压蒸馏时要小火加热，并注意观察温度计和压力表的读数。

• **Helpful Hints**

1. All the glassware used for the Claisen condensation reaction must be dried thoroughly and the reagents must be absolutely anhydrous[⑩].

2. The sodium metal to be used should be cut into small pieces to accelerate the reaction. Sodium metal is easy to be oxidized or reacts with moisture in the air. Therefore, the process of cutting and weighing of the sodium metal should be carried out rapidly[⑪].

3. The Claisen condensation reaction usually starts quickly. If the reaction is too vigorous, immerse the reaction flask in a cold-water bath to cool down the reaction mixture. If the reaction does not occur after a moment, heat the reaction mixture gently to promote the reaction.

4. 50% acetic acid should be added slowly and carefully because it reacts vigorously with unreacted sodium. Furthermore, acetic acid cannot be added in excess, so add it until the solution is slightly acidic (pH = 5-6). Otherwise, excess acetic acid will increase the solubility of the product in the aqueous layer, thus reducing the yield of acetoacetate.

• **Questions**

1. In the work-up procedure, what is the purpose of adding 50% acetic acid? Why should acetic acid be added carefully and slowly?

2. What is the purpose of each washing and extracting in this experiment?
(1) Washing the reaction mixture with saturated aqueous sodium chloride solution.
(2) Extracting the aqueous layer with 5 mL of ethyl acetate.
(3) Washing the organic layer with 5% sodium carbonate solution.

3. Discuss the differences observed in the IR and ^1H NMR spectrum of ethyl acetate and ethyl acetate that are consistent with the conversion in this experiment.

4. Mixed Claisen condensation reactions of two different esters can be successfully performed only when one of the two ester components has no α-H and thus fails to form an enolate ion. Predicate the major products of the following reactions.

$$(1)\ C_6H_5\overset{O}{\overset{\|}{C}}OEt + CH_3CH_2\overset{O}{\overset{\|}{C}}OEt \xrightarrow[2.\ CH_3COOH]{1.\ NaOC_2H_5/C_2H_5OH}$$

$$(2)\ CH_3\overset{O}{\overset{\|}{C}}OEt + EtO\overset{O}{\overset{\|}{C}}-\overset{O}{\overset{\|}{C}}OEt \xrightarrow[2.\ CH_3COOH]{1.\ NaOC_2H_5/C_2H_5OH}$$

Exp.5 Synthesis and Characterization of Aspirin

• **Objectives**

1. To study the principle and method of preparing Aspirin by acylation of salicylic acid.

---

⑩ 反应所用仪器要严格干燥，所用试剂必须绝对无水。

⑪ 实验中所用金属钠要切成小碎片。金属钠极易被氧化或和空气中的水反应，切割和称量金属钠时动作要快，不要接触水。

2. To practice vacuum filtration and recrystallization.

3. To learn how to characterize an organic compound by TLC analysis and spectroscopy.

• **Principle**

Since Hoffman synthesized aspirin (acetylsalicylic acid) in 1897, aspirin has been widely used to alleviate pains and treat a fever, and now it is still one of the largest selling analgesics. In recent years, more and more applications of aspirin have been discovered, such as treating immune diseases, preventing the occurrence of cerebrovascular diseases, minimizing the platelet aggregation, and slowing down the progression of the Alzheimer's disease, etc.

Aspirin is synthesized by the acylation reaction of salicylic acid using acetic anhydride as the acylation reagent. In the acylation reaction of salicylic acid, acetic anhydride preferentially reacts with the phenolic hydroxyl group rather than the carboxylate group. The reaction is demonstrated below.

$$\text{Salicylic acid} + (CH_3CO)_2O \xrightarrow{H^+} \text{Acetylsalicylic acid} + CH_3COOH$$

Salicylic acid is a bifunctional compound and can form intramolecular hydrogen bond, which hinders the acylation of the phenolic hydroxyl group. Thus, the direct reaction between salicylic acid and acetic anhydride can only take place at high temperature (150-160 ℃) to produce acetylsalicylic acid⑫. Besides, this reaction will work better when a small amount of concentrated acid, such as sulfuric acid or phosphoric acid, is used as the catalyst. This is because the acidic catalyst can destroy the intramolecular hydrogen bonding, thus allowing the reaction to proceed at lower temperature (60-80 ℃) and minimize the formation of the following polymers as by-products⑬.

In order to remove the polymers, the crude acetylsalicylic acid is firstly converted into its water-soluble salt, which can be easily separated from the water-insoluble polymers by vacuum filtration, so that the pure acetylsalicylic acid will be obtained by acidifying the resulting aqueous solution of the salt.

---

⑫ 水杨酸是一个双官能团化合物，能形成分子内氢键，这会阻碍酚羟基酰基化。所以，水杨酸与酸酐直接作用生成乙酰水杨酸的反应往往需要加在比较高的温度（150～160 ℃）方可完成。

⑬ 酸性催化剂的加入，可以破坏分子内氢键，使酰基化反应可以在较低温度（60～80℃）完成，反应温度的降低也会大大减少副产物的生成。

- **Apparatus**

Apparatuses required include 100 mL Erlenmeyer flask, Allihn condenser, Büchner funnel, filter flask, test tube and beaker.

- **Setting Up**

Place 2 g of salicylic acid and 5 mL of acetic anhydride in a dry 100 mL Erlenmeyer flask that contains several zeolites. Add 5 drops of concentrated phosphoric acid carefully to the flask while swirling the flask. Assemble the apparatus for reflux as shown in Figure 4.5.

Figure 4.5 Apparatus for the preparation of Asprin

- **Experimental Procedures**

1. Heat the reaction mixture gently for 15 min while swirling the flask occasionally, with the reaction temperature kept around 80 ℃ ⑭. The solids will be dissolved during this process.

2. Stop heating and allow the mixture to cool slightly.

3. Slowly pour the mixture into a beaker containing 50 mL of water while stirring. Then, allow the beaker to cool down in an ice-water bath to precipitate the crude aspirin.

4. Collect the solid by vacuum filtration and wash it thoroughly with several small portions of ice-cold water.

5. Transfer the crude aspirin to a beaker, and then add 25 mL of saturated solution of sodium carbonate to the beaker until no further gas is released and the pH of the obtained solution is about 8 ⑮.

6. Filter the solution in vacuum to remove the water-insoluble by-products. Wash the filtered cake further with 10 mL of water.

7. Transfer the resulting filtrate into a beaker and slowly add concentrated hydrochloric acid to the solution until the pH of the solution is 2.

8. Cool down the resulting solution in an ice-water bath to precipitate the crystal completely.

9. Collect the crystal by vacuum filtration, and wash it with several small portions of water and air-dried.

10. Further purify aspirin by recrystallization with ethyl acetate ⑯.

11. Weigh the dried product and calculate the percentage yield of acetylsalicylic acid.

- **Characterization of Aspirin**

1. Melting point: measure the melting point of the obtained dry acetylsalicylic acid. The reported melting point of acetylsalicylic acid is 138-140 ℃.

2. TLC analysis: dissolve 2 mg of acetylsalicylic acid and 2 mg of salicylic acid in 0.2 mL

---

⑭ 反应温度的控制十分重要。如果反应温度过高，副产物会大大增加，给后处理过程带来麻烦。

⑮ 为了除去反应过程中生成的少量聚合物，可先将粗产物乙酰水杨酸溶解在饱和碳酸氢钠溶液中，转化成水溶性的钠盐，然后利用聚合物不溶于水的性质，通过减压过滤将他们分开。

⑯ 粗乙酰水杨酸重结晶提纯时，不宜长时间加热，以免乙酰水杨酸发生水解。

of ethanol respectively. Load the two samples onto a TLC plate and develop the plate using a mixture of 1 : 1 ethyl acetate/petroleum ether as the developing solvent. After completing the development, visualize the plate under UV light at 254 nm and calculate the R_f values of the two components.

3. Infrared spectroscopy: record and interpret the IR spectrum of acetylsalicylic acid as a KBr pellet.

4. ^{1}H NMR spectroscopy: dissolve about 50 mg of acetylsalicylic acid in deuterated DMSO. Record and interpret the ^{1}H NMR spectrum of acetylsalicylic acid.

5. Ferric test: place a few crystals of acetylsalicylic acid into a test tube. Then, add 5 mL of distilled water to the tube and swirl the tube until the crystals are dissolved. Add 10 drops of 1% aqueous ferric trichloride solution to the tube and record the observation.

- **Helpful Hints**

1. All the glassware used for the acylation of salicylic acid should be dried completely[⑰].

2. Acetic anhydride is irritant. Handle it in the hood and do not breathe the fumes.

3. Phosphoric acid is corrosive. Wear gloves while handling this reagent.

4. Concentrated hydrochloric acid is a corrosive liquid, so wear latex gloves while handling it and thoroughly wash any areas of your skin that may come in contact with it.

5. When the acylation of salicylic acid is carried out at higher temperature, the following polymers are easily formed as by-products. Therefore, the reaction temperature must be controlled at around 80 ℃.

6. Either sodium carbonate solution or concentrated hydrochloric acid should be added slowly and carefully to avoid the occurrence of vigorous reaction, which may cause the loss of desired product[⑱].

7. Phenolic hydroxyl group can form color complex with aqueous ferric trichloride solution[⑲].

- **Questions**

1. Explain why the acetylation of salicylic acid with acetic anhydride occurs at the phenolic hydroxyl group rather than the carboxyl group?

2. What is the purpose of adding concentrated phosphoric acid in the acylation reaction of salicylic acid with acetic anhydride?

3. What measures should be taken to remove the by-products generated in this reaction?

---

⑰ 水杨酸的乙酰化反应所用的仪器必须绝对干燥，水杨酸使用前应在烘箱内105℃干燥1h，乙酸酐要重新蒸馏。

⑱ 向反应混合物中加饱和碳酸氢钠溶液或浓盐酸时，要小心并缓慢滴加，以免反应过于剧烈，造成产物损失。

⑲ 酚羟基可以与三氯化铁溶液发生显色反应。

4. Predict the acetylation products of the following two compounds with acetic anhydride under basic condition.

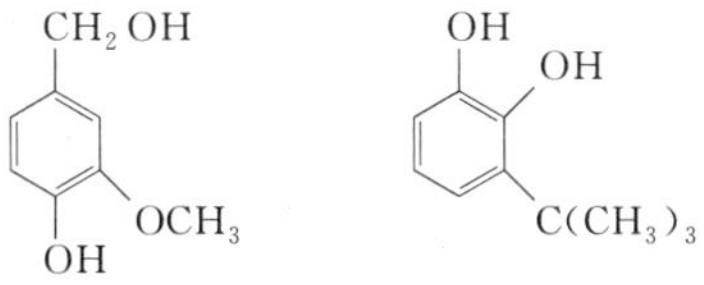

[Structures: 4-hydroxy-3-methoxybenzyl alcohol and 3-tert-butylcatechol]

5. Discuss the differences observed in the IR and ^{1}H NMR spectra of salicylic acid and acetylsalicylic acid.

Exp.6 Multi-step Synthesis of Sulfanilamide[50],[51]

• **Objectives**

1. To study the principle and method for multi-step synthesis of sulfanilamide.
2. To understand the mechanism of each synthesis step.
3. To practice recrystallization and vacuum filtration.

• **Principle**

The broad-spectrum antibacterial activity of sulfanilamide was first revealed in the 1930s by accident. Up to now, over one thousand derivatives of sulfanilamide have been synthesized using sulfanilamide as the precursor, and some of them are known as sulfa drugs. The structures of these sulfa drugs are rather similar, differing only in the groups bonded to the sulfonyl group.

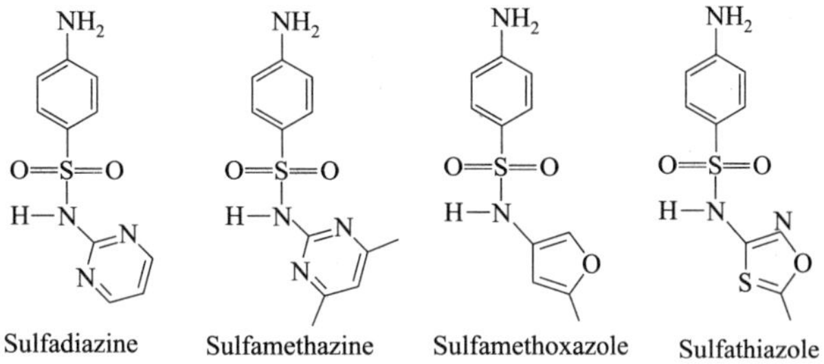

[Structures of Sulfadiazine, Sulfamethazine, Sulfamethoxazole, Sulfathiazole]

The preparation of sulfanilamide from acetanilide is an excellent example of a multi-step synthesis example, where the product of one step serves as the reactant for the subsequent step. Each of the intermediate compounds in the synthesis may be either isolated or used directly in the next step without further purification.

In this experiment, the aromatic ring of acetanilide is sulfonated by chlorosulfonic acid to produce 4-acetamidobenzenesulfonyl chloride, which then reacts with ammonia to

---

[50] Reference: Allen M S, Barbara A G, Melvin L D. Microscale and Miniscale Organic Chemistry Laboratory Experiments [M]. 2nd ed. McGraw-Hill press, 2004: 592-600.

[51] Reference: John C G, Stephen F M. Experimental Organic Chemistry: A Miniscale and Microscale Approach [M]. 5th ed. Cengage Learning press, 2011: 705-725.

yield benzene sulfonamide. Finally, the acetamido is hydrolyzed to form the target molecule sulfanilamide. The synthetic route of sulfanilamide from acetanilide is summarized below.

$$\text{Acetanilide} \xrightarrow{ClSO_3H} \text{4-NHCOCH}_3\text{-C}_6\text{H}_4\text{-SO}_2Cl \xrightarrow{NH_3} \text{4-NHCOCH}_3\text{-C}_6\text{H}_4\text{-SO}_2NH_2 \xrightarrow[Na_2CO_3]{HCl/H_2O} \text{Sulfanilamide}$$

1. Preparation of 4-acetamidobenzenesulfonyl chloride

The sulfonyl group, $-SO_2Cl$, can be introduced to the benzene ring by electrophilic substitution reaction of acetanilide and chlorosulfonic acid. Two equivalents of chlorosulfonic acid are required for this conversion: the first produces the electrophile SO_3 in situ to yield 4-acetamidosulfonic acid, while the second converts sulfonic acid to sulfonyl chloride⑱.

$$HOSO_2Cl \rightleftharpoons SO_3 + HCl$$

$$\text{PhNHCOCH}_3 \xrightarrow{ClSO_3H} \text{4-NHCOCH}_3\text{-C}_6\text{H}_4\text{-SO}_3H \xrightarrow{ClSO_3H} \text{4-NHCOCH}_3\text{-C}_6\text{H}_4\text{-SO}_2Cl$$

Although sulfonyl chlorides are not as moisture sensitive as carboxylic acid chlorides, they will eventually hydrolyze to sulfonic acids if exposed to the air for a period of time. Therefore, 4-acetamidobenzenesulfonyl chloride should be used immediately after being synthesized and does not need to be dried and purified further in this sequence⑲.

2. Reaction of 4-acetamidobenzenesulfonyl chloride with ammonia

This is the key step for the preparation of sulfa drugs, which involves nucleophilic addition of excess aqueous ammonia to 4-acetamidobenzenesulfonyl chloride, followed by the loss of HCl to form the 4-acetamidobenzenesulfonamide.

$$\text{4-NHCOCH}_3\text{-C}_6\text{H}_4\text{-SO}_2Cl + 2NH_3 \longrightarrow \text{4-NHCOCH}_3\text{-C}_6\text{H}_4\text{-SO}_2NH_2 + NH_4Cl$$

3. Hydrolysis of the acetamido group

After the sulfonamide group is introduced into the aromatic ring, the acetyl protecting

---

⑱ 芳基磺酰氯一般是通过芳烃的氯磺化反应来制备的。整个转化过程需要两当量的氯磺酸，一当量用来产生亲电试剂三氧化硫，并与芳烃发生取代反应生成中间体芳基磺酸。芳基磺酸接着会与另一当量的氯磺酸作用得到芳基磺酰氯。

⑲ 4-乙酰胺基苯磺酰氯一经合成应尽快使用，不得长时间放置，以免其发生水解反应。因为下一步 4-乙酰胺基苯磺酰氯与氨作用合成 4-乙酰胺基苯磺酰胺的反应是在水溶液中进行的，所以本实验得到的 4-乙酰胺基苯磺酰氯不需要干燥或进一步提纯。

group needs to be hydrolyzed to yield the free amino group. Amides can be hydrolyzed in either acidic or basic conditions while the sulfonamide group is not affected. An acid-catalyzed hydrolysis is performed in this experiment. The free amino group obtained after acidic hydrolysis will form a salt with the acid and the final product 4-aminoobenzenesulfonamide is obtained after neutralizing the aqueous solutions of the salt by a base[⑭].

$$\text{4-NHCOCH}_3\text{-C}_6\text{H}_4\text{-SO}_2\text{NH}_2 \xrightarrow[\text{2. NaHCO}_3]{\text{1. HCl/H}_2\text{O}} \text{4-NH}_2\text{-C}_6\text{H}_4\text{-SO}_2\text{NH}_2$$

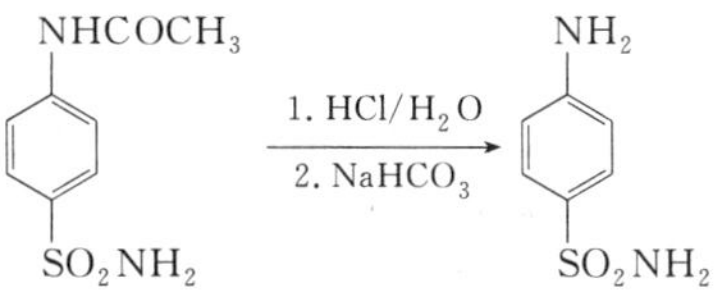

- **Apparatus**

Apparatuses required include 100 mL Erlenmeyer flask, Allihn condenser, long-necked glass funnel, beaker, Büchner funnel, filter flask, graduate cylinder, 250 mL round-bottom flask and separatory funnel.

- **Experimental Procedures**

1. Preparation of 4-acetamidobenzenesulfonyl chloride

(1) Add 5 g of acetanilide into a dry 100 mL Erlenmeyer flask. Place the flask on a hot plate and heat until the acetanilide is melted.

(2) Remove the reaction flask from the hot plate and allow the liquid to solidify, and then cool down it in an ice-water bath[⑮].

(3) Add 12.5 mL of chlorosulfonic acid via graduated cylinder at a time and immediately connect the gas-trap to the reaction flask[⑯].

(4) Let the flask stand at room temperature and shake it slightly until the solid disappears. Then, gently heat the flask in a warm-water bath for 10 min until no more hydrogen chloride gas is produced.

(5) Allow the flask to cool down to room temperature. Slowly and carefully pour the solution into 75 mL of ice water in a beaker while stirring vigorously[⑰].

(6) Wash the reaction flask with about 10 mL of ice water and combine the washings to the beaker.

(7) Stir the resulting solution for several minutes and allow the product to precipitate.

(8) Collect the solid by vacuum filtration and wash it with cold water.

(9) Press the solid with a spatula under vacuum to remove as much water as possible.

The obtained solid is 4-acetamidobenzenesulfonyl chloride and can be used immediately in the next reaction.

---

⑭ 酰胺基酸性条件下水解生成的氨基会立即与体系中的酸形成盐，用碱中和后即可得到最终产物 4-氨基苯磺酰胺。

⑮ 氯磺化反应十分剧烈，很难控制。将乙酰苯胺凝结后再滴加氯磺酸，是为了确保反应能平稳进行。

⑯ 一次性加完氯磺酸后，应马上连接上气体吸收装置，以免生成的氯化氢气体扩散，造成窒息。

⑰ 体系中残留的氯磺酸遇水会剧烈反应，放出热量。为了防止局部过热，造成 4-乙酰胺基苯磺酰氯的水解，这步操作要十分小心，应将反应液慢慢倒入冰水中，并充分搅拌。

2. Preparation of 4-acetamidobenzenesulfonamide

(1) Transfer the crude 4-acetamidobenzenesulfonyl chloride obtained in the previous reaction to a 100 mL Erlenmeyer flask.

(2) Slowly add 15 mL of concentrated ammonium to the flask while stirring⑱. White paste is produced during this process.

(3) Upon the completion of the addition, continue to stir the mixture for 10 min to complete the reaction.

(4) Add 10 mL of water to the resulting mixture and heat the mixture at 70-80 ℃ for about 0.5 h⑲. The resulting mixture can be used directly in the next reaction.

3. Hydrolysis of the acetamido group to form sulfanilamide

(1) Transfer the obtained 4-acetamidobenzenesulfonamide to a round-bottom flask, add 5 mL of concentrated hydrochloric acid and 10 mL of water to the flask⑳.

(2) Gently heat the mixture to reflux for 50 min while stirring.

(3) Stop heating. Add a small amount of activated carbon to the solution and then heat the mixture to boil for about 3-5 min.

(4) Filter the mixture when it is hot and transfer the filtrate to a beaker.

(5) Slowly add solid sodium bicarbonate to the filtrate while stirring until the solution is almost neutral㉑.

(6) Carefully add saturated sodium bicarbonate solution to the solution until it is just neutral㉒.

(7) Cool down the mixture in an ice-water bath to precipitate the solid completely.

(8) Collect the solid by vacuum filtration, and wash it with a small amount of cold water and air-dried.

(9) Purify the crude sulfanilamide by recrystallization from hot water.

(10) Weigh the dried product and calculate the percentage yield of sulfanilamide.

- **Helpful Hints**

1. Preparation of 4-acetamidobenzenesulfonyl chloride

(1) Make sure that all the glassware used in this reaction is dry, and be certain that the joints of the glassware are mated. Otherwise, noxious gases will escape.

(2) Chlorosulfonic acid is very corrosive. Wear double-layered gloves while working with

---

⑱ 此步反应搅拌一定要充分，否则会有一些未反应的 4-乙酰胺基苯磺酰氯包裹在产物 4-乙酰胺基苯磺酰胺里，影响下一步的水解反应。

⑲ 加热是为了除去多余的氨。

⑳ 由于溶液中氨的含量可能不同，有时加 5 mL 盐酸进行下一步水解反应可能不够。因而，在加热回流至固体完全消失后，应测一下溶液的酸碱性，如果溶液不呈现酸性，则需补加盐酸呈酸性后再继续加热。

㉑ 中和反应会放出大量的热，同时生成二氧化碳。因此要慢慢加入固体碳酸氢钠，并不断搅拌，以防止产物溢出。

㉒ 产物 4-氨基苯磺酰胺可溶在碱中，因而必须严格控制饱和碳酸氢钠溶液的用量并慢慢滴加，加至溶液恰好呈中性即可。

this noxious reagent and cleaning glassware containing residual amounts of this acid. Chlorosulfonic acid is a very strong acid and can cause severe burns. If this reagent spills on your skin, immediately flood the affected area with cold water and then rinse it with 5% sodium bicarbonate solution.

(3) Chlorosulfonic acid reacts violently with water, even with moisture in the air. Thus, the measuring and transferring of this reagent should be completed quickly in the hood. While cleaning the glassware that contains traces of chlorosulfonic acid, add cracked ice to the glassware and let the glassware remain there until the ice is melted. Then, wear gloves and rinse the glassware with copious amounts of water[⑮].

(4) The chlorosulfonation reaction is vigorous and difficult to control. Coagulate acetanilide first before the addition of chlorosulfonic acid, so that the reaction can proceed smoothly.

(5) After the chlorosulfonation is completed, the reaction mixture should be poured into ice water slowly to prevent hydrolysis of 4-acetamidobenzenesulfonyl chloride caused by local overheating. The remaining chlorosulfonic acid in the mixture can react with water vigorously, so the addition should be performed carefully.

(6) Given that sulfonyl chloride will hydrolyze to sulfonic acid if being stored in the air for a long time, use 4-acetamidobenzenesulfonyl chloride immediately after it is isolated.

2. Preparation of 4-acetamidobenzenesulfonamide

(1) Concentrated ammonium is caustic and may cause burns. Wear latex gloves while handling or transferring this reagent, and use it in a hood. If any of the liquid comes in contact with your skin, immediately flood the affected area with cold water.

(2) This reaction of 4-acetamidobenzenesulfonyl chloride with ammonia is the conversion of a solid compound into another solid one, making it necessary to stir the reaction mixture thoroughly during the reaction process. Otherwise, some unreacted 4-acetamidobenzenesulfonyl chloride may be packaged in the final product, which will affect the purity of the 4-acetamidobenzenesulfonamide.

3. Hydrolysis of the acetamido group to form sulfanilamide

(1) Concentrated hydrochloric acid is corrosive and may cause serious burns. Wear latex gloves while handling or transferring this reagent. If any acid comes in contact with your skin, immediately flood the affected area with cold water and then rinse it with 5% sodium bicarbonate solution.

(2) A large amount of carbon dioxide will be released during the neutralization step, so the solid sodium bicarbonate should be added slowly while stirring to prevent the product from overflowing.

---

⑮ 氯磺酸遇水剧烈反应，甚至与空气中的湿气也会发生剧烈反应。因此，该试剂的量取应十分小心且动作要快，并且要在通风橱中进行。在清洗残留有氯磺酸的玻璃器皿时，可先在玻璃器皿中加入碎冰，等到冰融化后，再用大量的水冲洗玻璃器皿。

(3) Sulfanilamide is soluble in the excess base, so the amount of sodium bicarbonate must be carefully controlled during neutralization.

• **Questions**

1. Explain why 4-acetamidobenzenesulfonyl chloride is much less susceptible to hydrolysis than 4-acetamidobenzoyl chloride.

$$CH_3COHN-\underset{\text{4-Acetamidobenzoyl chloride}}{\underline{}}-COCl$$

2. Both 2-Aminopyridine and 3-aminopyridine contain two nitrogen atoms. When 4-acetamidobenzenesulfonyl chloride reacts with them, why does the amino group react rather than the pyridine nitrogen?

2-Aminopyridine   3-Aminopyridine

3. Can solid sodium hydroxide be used instead of solid sodium carbonate in the final neutralization step?

4. Outline a possible synthesis of the sulfanilamide derivative of the following compound, using benzene as the only source of an aromatic ring.

$$H_2N-\underset{}{\underline{}}-SO_2NH-\underset{}{\underline{}}$$

5. Discuss the differences observed in the IR and ^1H NMR spectra of 4-acetamido-benzenesulfonyl chloride and 4-acetamidobenzenesulfonamide that are consistent with the formation of the latter in this procedure.

Exp.7 A Solvent Free Cannizzaro Reaction and Thin-layer Chromatography[①]

• **Objectives**

1. To understand the mechanism of Cannizzaro reaction.
2. To learn to carry out a reaction under grinding conditions.
3. To practice characterizing an organic compound by TLC analysis.
4. To practice acidification and vacuum filtration.

• **Principle**

When an aldehyde with no α-H reacts with a strong base, the Cannizzaro reaction occurs and results in a simultaneous oxidation and reduction, or disproportionation. In the traditional Cannizzaro reaction, concentrated aqueous solution of a base, such as sodium hydroxide or po-

---

[①] Reference: Sonthi P, Bhinyo P. A Facile Solvent-Free Cannizzaro Reaction: An Instructional Model for Introductory Organic [J]. Journal of Chemical Education. 2009, 86 (1): 85-86.

tassium hydroxide is usually applied to provide a strong basic reaction condition. As the solubility of aldehydes in aqueous solutions is poor, the normal Cannizzaro reaction is heterogeneous, making either vigorous stirring or water-miscible organic solvent, such as THF or alcohols, necessarily important to facilitate the reaction. Meanwhile, high temperature is also required for the completion of the Cannizzaro reaction[165].

The study of organic reactions in the absence of solvent is more meaningful today to highlight green chemistry. Many works focusing on the solvent-free reactions have been published. The solvent-free reactions generally occur quickly, and can be conducted easily under convenient conditions, such as grinding, microwave irradiation, or mechanical stirring, in the presence or absence of heat. In addition, the solvent-free reactions also have the advantages of no organic solvent pollution, energy efficiency, environment-friendly and simple work-up procedure[166], which make them also attractive in undergraduate laboratories.

In this experiment, the solvent-free disproportionation of 2-chlorobenzaldehyde is undertaken under grinding conditions in the presence of potassium hydroxide (KOH). Along with the grinding process, 2-chlorobenzaldehyde and the solid base are mixed thoroughly to achieve the formation of the products 2-chlorobenzyl alcohol and 2-chlorobenzoate. These two products can be isolated easily according to their different solubilities in water, in which 2-chlorobenzyl alcohol insoluble in water is collected by vacuum filtration, while 2-chlorobenzoic acid will be precipitated out by acidifying the aqueous solution of 2-chlorobenzoate. Obviously, organic solvents are hereby eliminated not only in the initial reaction process but also in the later work-up procedure[167].

$$\text{2-ClC}_6\text{H}_4\text{CHO} \xrightarrow[\text{2. H}_3\text{O}^+]{\text{1. KOH, Grinding}} \text{2-ClC}_6\text{H}_4\text{CH}_2\text{OH} + \text{2-ClC}_6\text{H}_4\text{COOH}$$

- **Apparatus**

Apparatuses required include 50 mL beaker, Büchner funnel, filter flask, graduate cylinder, mortar, pestle, water bump and apparatus for magnetic stirrer.

- **Experimental Procedures**

1. Add 2 mL of 2-chlorobenzaldehyde and 1.5 g of potassium hydroxide into a mortar.
2. Grind the mixture continuously for about 30 min using a pestle.
3. Upon the completion of the reaction[168], add 10 mL of water to the resulting mixture and stir it thoroughly.
4. Filter the suspension and collect the solid by vacuum filtration.

---

[165] 由于醛类化合物在水溶液中的溶解度较小，所以坎尼扎罗反应一般是非均相反应。为了促进反应物之间的有效接触，反应通常在较高温度下进行，且需要剧烈搅拌或使用能和水互溶的有机溶剂，如四氢呋喃或醇等。

[166] 无溶剂反应一般可在研磨、微波或机械搅拌辅助下进行。无溶剂反应具有反应速率快、无有机溶剂污染、环境友好、能源效率高、后处理过程简单等优点。

[167] 研磨辅助的2-氯苯甲醛的歧化反应，无论是反应过程还是后处理过程都实现了无溶剂化。

[168] 反应结束时，反应物应该由黏性混合物变成浓稠的膏状，可以此作为反应的终点。

5. Wash the solid with 5 mL of water twice and air-dried as product 2-chlorobenzyl alcohol.

6. Combine all the filtrate and add concentrated hydrochloric acid slowly to the filtrate until the pH of the solution is 1.

7. Filter the resulting mixture and collect the solid by vacuum filtration.

8. Wash the solid thoroughly with water and air-dried as product 2-chlorobenzoic acid.

9. Weigh the two products and calculate their percentage yields respectively.

- **Characterization of the Products**

1. Melting point measurement of 2-chlorobenzyl alcohol and 2-chlorobenzoic acid.

2. TLC analysis follows the following steps:

(1) Dissolve 5 mg of both two products in 1 mL of ethanol respectively.

(2) Load the samples onto a TLC plate.

(3) Develop the plate using 1 : 1 mixture of ethyl acetate and petroleum ether as the developing solvent.

(4) Visualize the plate under UV light at 254 nm.

(5) Calculate the R_f values on the TLC and compare the values with those of the authentic samples.

- **Helpful Hints**

1. 2-Chlorobenzaldehyde, 2-chlorobenzoic acid, and 2-chlorobenzyl alcohol are all skin, eye, and respiratory irritants. Use them in a well-ventilated area and avoid contact or inhalation for a long period.

2. Potassium hydroxide is corrosive. Wear gloves while handling it and avoid skin contact with it. If potassium hydroxide contacts the skin, wash the affected area with copious amounts of cool water.

3. The progress of the disproportionation of 2-chlorobenzaldehyde can be monitored by thin-layer chromatography as follows: dissolve the reaction mixture in ethanol, spot on the TLC plate, and then develop the plate using the 1 : 1 mixture of ethyl acetate and petroleum ether as the developing solvent[169].

4. The melting point of 2-chlorobenzoic acid is 140-142 ℃, while that of 2-chlorobenzyl alcohol is 73-74 ℃.

- **Questions**

1. What are the advantages of solvent-free reactions compared to the traditional ones?

2. What is the basis for the separation and purification of the two products in this experiment?

3. Why is the melting point of 2-chlorobenzoic acid much higher than that of 2-chlorobenzyl alcohol?

---

[169] 可以用薄层色谱来追踪2-氯苯甲醛歧化反应的进程，确定反应的终点。在进行薄层色谱分析时，一般用乙醇为溶剂来溶解反应混合物，乙酸乙酯和石油醚的等量混合物作为展开剂来进行薄层色谱分析。

4. Propose a mechanism of the disproportionation of 2-chlorobenzaldehyde, showing the structures of the intermediates and using curved arrows to indicate the electron flow in each step.

5. Discuss the differences observed in the IR and ^{1}H NMR spectra of 2-chlorobenzaldehyde, 2-chlorobenzoic acid, and 2-chlorobenzyl alcohol that are consistent with the formation of the products from 2-chlorobenzaldehyde by the Cannizzaro reaction.

Exp.8 Preparation of 4-Vinylbenzoic Acid by a Wittig Reaction[①]

• **Objectives**

1. To study the principle and method of preparing an alkene by a Wittig reaction.
2. To understand the mechanism of the Wittig reaction.
3. To practice vacuum filtration, washing and recrystallization.
4. To learn to characterize an organic compound by melting point measurement and TLC analysis.

• **Principle**

Wittig reaction is one of the best known and useful method to construct C=C bond, starting with an aldehyde or ketone and a triphenylphosphorus ylide ($R_2C=PPh_3$), also called Wittig reagent. In the Wittig reaction, a triphenylphosphorus ylide is added to an aldehyde or ketone by a nucleophilic addition reaction to yield a dipolar intermediate, which undergoes ring closure to give a four-membered cyclic transition state, and will then be spontaneously decomposed to give an alkene plus triphenylphosphine oxide ($O=PPh_3$).

$$\text{Aldehyde/ketone} + Ph_3P=CHR \rightleftharpoons [\text{Betaine}] \rightleftharpoons [\text{Oxaphosphetane}] \longrightarrow \text{alkene} + Ph_3P=O$$

The net effect of the Wittig reaction is that the oxygen atom of the aldehyde or ketone and the $R_2C=$ bonded to phosphorus atom exchange their places. The Wittig synthesis of alkenes has two important advantages over other methods for formation of C=C bonds: (1) the Wittig reaction can be performed under mild conditions, and (2) the Wittig reaction always yields a pure alkene of predictable structure with no isomers (except E, Z isomers)[②].

In this experiment, 4-vinylbenzoic acid is synthesized via the Wittig reaction of formaldehyde with the corresponding ylide, which is obtained by treating 4-bromomethylbenzoic acid with triphenylphosphine in the presence of sodium hydroxide.

---

① Reference: 李英, 邵莺. Laboratory Experiments for Organic Chemistry (有机化学实验) [M]. 南京: 南京大学出版社, 2021: 196-200.

② 和其他构建碳碳双键的方法相比, 用Wittig反应来合成烯烃具有两大优点: (1) 反应条件缓和; (2) 得到的烯烃结构可预测, 不会产生异构现象。

$$\underset{\text{COOH}}{\text{C}_6\text{H}_4\text{-CH}_2\text{Br}} \xrightarrow[\text{Acetone}\triangle]{\text{Ph}_3\text{P}} \underset{\text{COOH}}{\text{C}_6\text{H}_4\text{-CH}_2\overset{+}{\text{PPh}_3}\text{Br}^-} \xrightarrow[\text{2. HCl}]{\text{1. HCHO, NaOH}} \underset{\text{COOH}}{\text{C}_6\text{H}_4\text{-CH=CH}_2}$$

- **Apparatus**

Apparatuses required include 100 mL round-bottom flask, Allihn condenser, Büchner funnel, filter flask, Erlenmeyer flask, addition funnel and apparatus for magnetic stirrer.

- **Preparation of the Phosphonium Salt**

1. Setting Up

Place 4.3 g of 4-bromomethylbenzoic acid and 60 mL of acetone in a 100 mL round-bottom flask. Mix the mixture thoroughly until the solid is dissolved. Add 5.2 g of triphenylphosphine in one portion to the flask, and assemble the reflux apparatus as shown in Figure 4.6.

2. Experimental Procedures

(1) Heat the reaction mixture to reflux for 45 min.

(2) Stop heating and allow the mixture to cool down to room temperature.

(3) Collect the precipitated solid by vacuum filtration, and wash the solid with diethyl ether (2×20 mL) and air-dried.

(4) Weigh the solid and use it directly for the subsequent step.

3. Helpful Hints

Figure 4.6 Apparatus for the preparation of the phosphonium salt

(1) Diethyl ether is a toxic and volatile organic solvent. Use it in a well-ventilated area and avoid contact or inhalation for a long period. Diethyl ether is also an inflammable liquid, so make sure that there are no flames in the vicinity while handling it.

(2) The obtained phosphonium salt can be used directly without further purification.

- **Preparation of 4-Vinylbenzoic Acid**

1. Setting Up

Place 3.7 g of phosphonium salt obtained in the previous step, 32 mL of 37% aqueous formaldehyde and 15 mL of water in a 100 mL Erlenmeyer flask. Mix the mixture thoroughly and assemble the apparatus as shown in Figure 4.7.

2. Experimental Procedures

(1) Make a solution by dissolving 2.5 g of sodium hydroxide in 15 mL of water, and then transfer the solution to the addition funnel.

(2) Slowly add the solution of sodium hydroxide to the mixture in the Erlenmeyer flask while stirring the reaction mixture vigorously❶.

(3) Upon the completion of the addition, continue to stir the resulting mixture at room temperature for another 45 min.

(4) Turn off the stirrer and filter the resulting slurry under vacuum.

Figure 4.7 Apparatus for the preparation of 4-vinylbenzoic acid

(5) Collect the solid and wash it with 10 mL of water.

(6) Combine all the filtrate and allow it to cool down in an ice-water bath.

(7) Carefully add concentrated hydrochloric acid to the filtrate to precipitate the desired product❷.

(8) Collect the crude product by vacuum filtration and wash it with 30 mL of cold water.

(9) Purify the crude product by recrystallization using 30% aqueous ethanol solution.

(10) Weigh the pure product and calculate the percentage yield of 4-vinylbenzoic acid.

3. Characterization of 4-Vinylbenzoic Acid

(1) Melting point measurement of 4-vinylbenzoic acid.

(2) TLC analysis in accordance with the following steps:

① Dissolve 2 mg of 4-vinylbenzoic acid and 4-bromomethylbenzoic acid in 1 mL of dichloromethane respectively.

② Load the samples onto a TLC plate.

③ Develop the plate using 1∶1 mixture of ethyl acetate and dichloromethane as the developing solvent.

④ Visualize the plate under UV light at 254 nm.

⑤ Calculate the R_f values on the TLC plate and compare the values with those of the authentic samples.

4. Helpful Hints

(1) The solution of sodium hydroxide is very caustic. Wear gloves while handling it. If sodium hydroxide comes in contact with your skin, wash the affected area with copious amounts of water.

(2) Concentrated hydrochloric acid fume is suffocative and corrosive, so handle it in the hood. If any acid spills on your skin, wash it off with large amount of water and then with dilute sodium bicarbonate solution.

(3) The addition of sodium hydroxide solution is to produce the ylide and to dissolve the

---

❶ 加入氢氧化钠溶液的目的有两个，一是将季鏻盐转化为相应的磷叶立德，二是将产物转化成相应的盐溶于水中，便于后面分离、提纯。

❷ 浓盐酸应慢慢滴加，以免反应过于剧烈造成产物外溢。

product 4-vinylbenzoic acid into the aqueous layer.

(4) The ylides for the Wittig reaction are usually prepared in the presence of a strong base, such as *n*-butyllithium. In this experiment, sodium hydroxide is enough to remove the α-H from benzylphosphonium halides to form the ylide due to the relatively acidity of this α-H (pK_a=20)[⑭].

(5) Concentrated hydrochloric acid should be added slowly while stirring to avoid spraying.

(6) The melting point of 4-vinylbenzoic acid is 142-144 ℃.

• **Questions**

1. Benzyltriphenylphosphonium bromide is prepared via the S_N2 reaction of triphenylphosphine and benzyl bromide. Write down an equation for this reaction and point out which is the nucleophile and which is the electrophile in this transformation?

2. Triphenylphosphine oxide is a by-product of the Wittig reaction. Which step in this experiment is this compound removed from the product?

3. Which of the following compounds can be used to prepare Wittig reagents? Explain your answer.

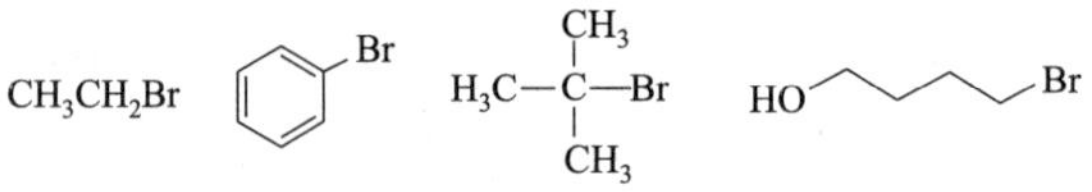

4. Devise an efficient synthesis route for each of the following alkenes by Wittig reactions.

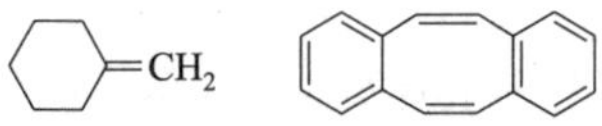

5. Wittig reagents react easily with aldehydes or ketones but slowly or not at all with esters. Explain the reason.

Exp.9 Preparation of Dimedone via the Robinson Annulation[⑮]

• **Objectives**

1. To understand the mechanism of Robinson Annulation.
2. To learn to perform a reaction under anhydrous and refluxing condition.
3. To practice the extraction and distillation of a liquid compound.
4. To practice vacuum filtration and recrystallization.

---

⑭ Wittig 反应的原料之一磷叶立德通常是在强碱，如正丁基锂存在下制备的。在本实验中，由于苄基季鳞盐中 α-H 的酸性较强（pK_a=20），因而氢氧化钠足以将 α-H 除去生成相应的磷叶立德。

⑮ Reference：李英，邵莺. Laboratory Experiments for Organic Chemistry（有机化学实验）[M]. 南京：南京大学出版社，2021：205-208.

• Principle

The Robinson annulation is a two-step process that combines a Michael addition with an intramolecular carbonyl condensation reaction such as the aldol condensation or Dieckmann condensation reaction. It is extensively used in organic synthesis to create a six-membered ring by forming three new carbon-carbon bonds[16]. For instance, in this experiment, dimedone is synthesized by the Robinson annulation of diethyl malonate with 4-methylpent-3-en-2-one in the presence of sodium ethoxide.

This reaction begins with the treatment of diethyl malonate, an active methlyene compound, with sodium ethoxide to produce a nucleophilic enolate. This enolate then does a 1,4 addition to 4-methylpent-3-en-2-one to achieve a 1,5-dicarbonyl compound. Subsequently, the α-hydrogen of the ketone residue in the 1,5-dicarbonyl compound is removed by a large excess of base. Meanwhile, a Dieckmann condensation takes place and a six-membered ring is constructed. The final product dimedone is obtained followed by a basic hydrolysis and an acid treated decarboxylation.

• Apparatus

Apparatuses required include 100 mL three-necked flask, Allihn condenser, Büchner funnel, filter flask, separatory funnel, addition funnel, 100 mL round-bottom flask, distillation head, Liebig condenser, thermometer, thermometer adapter, receiver adapter, Erlenmeyer flask, graduate cylinder and apparatus for mechanical stirrer.

• Setting Up

Place 16 mL of 25% sodium ethoxide in anhydrous ethanol into an oven-dried three-necked flask and assemble the apparatus as shown in Figure 4.8.

Figure 4.8 Apparatus for the preparation of dimedone

• Experimental Procedures

1. Place 7.6 mL of diethyl malonate in the addition funnel and add it dropwise to the reaction mixture in the flask over 5 min while stirring.

2. Add 5 mL of anhydrous ethanol[17] to rinse the addition funnel into the flask.

3. Add 5 mL of 4-methylpent-3-en-2-one dropwise to the reaction mixture through the addition funnel, and then add another 4 mL of anhydrous ethanol.

---

[16] 罗宾逊成环反应是一种重要的构筑六元环化合物的反应。它是由一个迈克尔加成反应和一个分子内羟醛缩合反应串联而成的反应。

[17] 罗宾逊成环反应所用反应试剂必须干燥。

4. Heat the mixture to reflux gently for 45 min while stirring⑯.

5. Slowly add the solution containing 6.3 g of sodium hydroxide in 25 mL of water to the resulting mixture through the addition funnel. Heat the mixture to reflux for another 45 min after the addition is completed⑰.

6. Allow the mixture in the flask to cool down slightly. Then, add several zeolites to the flask and assemble the flask for simple distillation.

7. Concentrate the reaction mixture by distilling out around 35 mL of the azeotrope of ethanol and water⑱.

8. Allow the residue in the flask to cool down in an ice-water bath and then transfer it into a separatory funnel.

9. Extract the cooled residue with 25 mL of diethyl ether⑲ and save the aqueous layer.

10. Wash the diethyl ether layer with 2×10 mL of water and combine the washings with the saved aqueous layer.

11. Acidify the combined aqueous solution with concentrated hydrochloric acid to pH=1 and heat the resulting solution to reflux for 15 min⑳.

12. Cool the mixture down in an ice-water bath to precipitate the crystals completely.

13. Collect the crystals by vacuum filtration and wash them with 25 mL of water, then 25 mL of petroleum ether㉑ and air-dried.

14. Purify the crude dimedone by recrystallization with water.

15. Weigh the pure product and calculate the percentage yield of dimedone.

• Helpful Hints

1. Sodium ethoxide is toxic irritant and caustic, so use it in a well-ventilated area and avoid inhalation. Wear gloves at all times. If sodium ethoxide comes in contact with your skin, wash the affected area with copious amounts of water.

2. Sodium hydroxide solution is caustic. Wear gloves at all times. If NaOH comes in contact with your skin, wash the affected area with copious amounts of water.

3. Concentrated hydrochloric acid fume is suffocative and corrosive, so handle it in the hood. If any acid spills on your skin, wash it off with large amount of water and then with dilute sodium bicarbonate solution.

4. All the glassware used in the initial reaction should be dried in the oven㉒.

5. Stirring is necessary for this reaction to ensure that the reaction mixture can be mixed completely.

6. Heating is required for the basic hydrolysis(saponification)and the acid treated decar-

---

⑯ 反应过程中必须搅拌，以确保反应物混合均匀，减少副反应的发生。
⑰ 酯基在碱性条件下的水解反应（皂化反应）需要在加热条件下方可进行。
⑱ 通过简单蒸馏，反应混合物中的乙醇与水以共沸物的形式蒸出，反应液得到浓缩。
⑲ 用乙醚萃取的目的是除去水相中残留的有机物。
⑳ 脱羧反应必须在加热的条件下进行，所以反应物酸化后需加热回流 15min。
㉑ 用石油醚洗涤产物的目的是使产物快速干燥。
㉒ 罗宾逊成环反应所用的仪器必须事先在烘箱中干燥。

boxylation.

7. The melting point of dimedone is 148-149 ℃.

- **Questions**

1. Propose the stepwise reaction mechanism of the dimedone formation via Robinson annulation of diethyl malonate with 4-methylpent-3-en-2-one followed by basic hydrolysis and acid treated decarboxylation.

2. Why should all the glassware used for the initial reaction be dried completely?

3. Which structural characteristics of carboxylic acids are easy to be decarboxylated while being heated?

4. Sodium ethoxide is used as a base to remove the α-H of diethyl malonate to form the enolate in the initial Michael addition. Can sodium hydroxide be used to replace sodium ethanol in this step?

5. What is the purpose of the extraction and washing operation in the work-up procedures of this experiment?

(1) Extracting the reaction mixture with 25 mL of diethyl ether.

(2) Washing the ether layer with water(2×10 mL).

Exp.10 Preparation of Cinnamic Acid by the Knoevenagel Condensation

- **Objectives**

1. To study the mechanism of Knoevenagel condensation reaction.
2. To practice the vacuum filtration and recrystallization of solid compounds.
3. To learn to precipitate solids by pH adjustment.
4. To practice melting point measurement with a digital melting-point apparatus.

- **Principle**

The Knoevenagel condensation is actually an aldol-like condensation reaction, in which aldehydes or ketones react with active methylene compounds, such as β-keto esters, malonic esters and malonodinitrile in the presence of a weak base, followed by hydrolysis or decarboxylation upon heating to give α,β-unsaturated carbonyl compounds[⑱]. The mechanism of the Knoevenagel condensation depends on the type of base used in the reaction. Weak bases such as amines, piperidine or pyridine as well as some inorganic bases are always used as the catalyst for this reaction. In order to get a clean condensation product and avoid competitive side reactions, there are generally no α-H, but an unhindered carbonyl group, in the aldehyde or ketone in Knoevenagel condensation reaction[⑲].

In this experiment, cinnamic acid is synthesized via the Knoevenagel condensation reac-

---

⑱ Knoevenagel 缩合反应是指具有活性亚甲基的化合物在弱碱性物质的催化作用下，与醛、酮发生类似于羟醛缩合的反应，经脱水、脱羧而得到 α,β-不饱和羰基化合物或其相关化合物的反应。

⑲ 为了得到单一的缩合产物，避免副反应的发生，Knoevenagel 缩合反应底物之一的醛或酮一般不含有 α-H，但含有一个空间位阻小、活性高的羰基。

tion of benzaldehyde with malonic acid in the presence of a mixture of pyridine and piperidine, followed by the decarboxylation upon heating. This reaction can be accelerated by removing the formed side product of water using methods including azeotropic distillation or addition of a dehydration reagents.

$$\text{C}_6\text{H}_5\text{CHO} + \text{CH}_2(\text{COOH})_2 \xrightarrow[\text{H}_3\text{O}^+, \triangle]{\text{Pyridine, Piperidine}} \text{C}_6\text{H}_5\text{CH=CHCOOH}$$

- **Apparatus**

Apparatuses required include 50 mL round-bottom flask, Allihn condenser, Büchner funnel, filter flask, separatory funnel, and graduate cylinders.

- **Setting Up**

Place 3.1 g of malonic acid and 5 mL of pyridine into a 50 mL round-bottom flask. Mix the reaction mixture thoroughly and assemble the flask for refluxing apparatus as shown in Figure 4.9.

- **Experimental Procedures**

1. Add 20 mL of water, 8 g of potassium carbonate and 3.6 mL of benzaldehyde to a separatory funnel, and shake the funnel well to mix the mixture thoroughly[⑰].

2. Stand the separatory funnel for 0.5 h until the content in the funnel is separated into two distinct layers. Then, discard the aqueous layer and transfer the organic layer to the reaction flask, which is pre-warmed in a water-bath at 60℃.

3. Add 10 drops of piperidine to the flask and then heat the resulting mixture to 95℃[⑱].

4. Stop heating when no bubbles can be observed.

5. Add 40 mL of 2 mol/L hydrochloric acid to the reaction mixture and allow the solution to cool down to room temperature[⑲].

6. Collect the solid by vacuum filtration and further wash it sequentially with 3×20 mL of 2 mol/L hydrochloric acid, 3×20 mL of water, 3×10 mL of 20% ethanol and 3×20 mL of petroleum ether in turn.

7. Dry the solid in an oven at 80℃.

8. Purify the crude product by recrystallization with $5:1 (V/V)$ of water and ethanol.

9. Weigh the product and calculate the percentage yield of cinnamic acid.

Figure 4.9 Apparatus for the preparation of cinnamic acid

---

⑰ 此步操作的目的是提纯苯甲醛,除去苯甲醛中所含的少量杂质苯甲酸。

⑱ 要控制反应的温度在95℃左右。温度过低,缩合反应难以进行。温度过高,副反应增加,影响产品的收率和纯度。

⑲ 此步加入盐酸的目的是发生淬火反应,且使产物析出。

- **Characterization of Cinnamic Acid**

1. Melting point measurement of cinnamic acid.

2. TLC analysis using a mixture of 90∶10(V/V) dichloromethane and methanol as the developing solvent.

- **Helpful Hints**

1. Benzaldehyde is toxic and irritating. Wear gloves and avoid skin contact while handling it.

2. Benzaldehyde having been stored for a long time will be easily oxidized to form benzoic acid and the acid formed can be removed by being treated with a base.

3. Pyridine and piperidine are irritating to the skin, eyes and mucous membranes. Wear gloves and avoid skin contact while handling them. Perform this reaction in the hood to minimize exposure or inhalation.

4. Pyridine acts as both the solvent and the condensation agent in this reaction[190].

5. The amount of piperidine added should not be excessive. Otherwise, the solution will be more basic, which may cause some side reactions, such as disproportionation of benzaldehyde or its own polymerization[191].

6. The addition of 2 mol/L hydrochloric acid to the reaction mixture is to quench the reaction and precipitate the product.

7. The final product of cinnamic acid might undergo polymerization[192].

8. Cinnamic acid has two *cis*- and *trans*- isomers, and the product prepared in this experiment is the *trans*-isomer with the melting point of 135 ℃, *trans*-cinnamic acid is an irritant. Wear gloves and avoid skin contact.

- **Questions**

1. Symbolize the flow of electrons using curved arrows, and write down the stepwise mechanism of the Knoevenagel condensation reaction of benzaldehyde with malonic acid in the presence of a mixture of pyridine and piperidine.

2. Explain why piperidine and pyridine (rather than sodium hydroxide) are used as the basic catalyst in the Knoevenagel condensation reaction of benzaldehyde with malonic acid.

3. Why do some bubbles generate when the reaction mixture is heated to 80 ℃?

4. What is the purpose of the following washing operation in the workup procedures of this experiment?

(1) Washing the solid with 3×20 mL of 2 mol/L hydrochloric acid.

(2) Washing the solid with 3×20 mL of water.

(3) Washing the solid with 3×10 mL of 20% ethanol.

(4) Washing the solid with 3×20 mL of petroleum ether.

---

[190] 吡啶在反应中既是溶剂又是缩合剂。

[191] 苯甲醛在浓碱条件下易发生歧化和自身氧化等副反应，导致产物肉桂酸颜色变深，纯度降低。因此要控制哌啶的加入量，不宜过量。

[192] 产物肉桂酸会发生聚合反应。

5. Library project: Find the other methods in the literature of synthesizing cinnamic acid and illustrate the experimental procedures of each method.

Exp.11　Microwave-assisted Preparation[193] of Benzoic Acid and Ethyl Benzoate

- **Objectives**

1. To study the principle and method of preparing benzoic acid by microwave-assisted oxidation reaction of benzyl alcohol.

2. To study the principle and method of preparing esters by Fischer esterification.

3. To practice the continuous removal of product water in esterification reaction using a water separator[194] (Dean-Stark apparatus).

4. To practice extraction, washing, drying and vacuum distillation.

- **Principle**

Due to the significant advantages of the microwave-assisted reaction, such as greatly shortened reaction time, reduced exposure to toxic chemicals, decreased thermal degradation[195], lower cost, simple work-up procedure and high yield of the product, etc., it is considered much more instructive compared with the conventional direct heating method. In recent years, microwave apparatus has been widely applied to organic synthesis. In this experiment, a microwave-assisted method is investigated to perform the oxidation of benzyl alcohol to benzoic acid. The reaction is shown as follows.

$$\text{PhCH}_2\text{OH} \xrightarrow[\text{Microwave}]{\text{KMnO}_4/\text{H}_2\text{O}} \text{PhCOOK} \xrightarrow{\text{H}^+} \text{PhCOOH}$$

An ester can be synthesized by a nucleophilic acyl substitution reaction[196] of a carboxylic acid with an alcohol using mineral acids as the catalyst. This kind of reaction is also called Fischer esterification[197]. For example, in this experiment, ethyl benzoate is synthesized by Fischer esterification of benzoic acid and ethanol in the presence of concentrated sulfuric acid.

$$\text{PhCOOH} + \text{CH}_3\text{CH}_2\text{OH} \xrightleftharpoons{\text{H}^+} \text{PhCOOCH}_2\text{CH}_3 + \text{H}_2\text{O}$$

Herein, sulfuric acid is used as a catalyst that protonates the carbonyl oxygen atom, thus activating the benzoic acid toward the nucleophilic attack by the ethanol. There always exists a reversible equilibrium in this reaction. Therefore, some special measures are therefore proposed to drive the equilibrium toward completion, which involves the application of a large excess of ethanol and the removal of the newly produced water out of the reaction mixture by

---

[193]　微波辅助合成。
[194]　分水器。
[195]　减少接触有毒化学品，减少热降解。
[196]　酰基的亲核取代
[197]　费歇尔酯化反应。

azeotropic distillation during the esterification using the Dean-Stark apparatus[48].

- **Apparatus**

Apparatuses required include 250 mL three-necked flask, microwave oven, Allihn condenser, Büchner funnel, filter flask, 100 mL round-bottom flask, water separator, separatory funnel, 50 mL round-bottom flask, distillation head, condenser, thermometer, thermometer adapter, receiver adapter, Erlenmeyer flask, beakers and apparatus for vacuum distillation.

- **Preparation of Benzoic Acid**

1. Setting Up

Place 4.0 g of potassium permanganate, 2.0 g of sodium carbonate and 30 mL of water into a 250 mL round-bottom flask. Mix the mixture evenly, and then add 2.1 mL of benzyl alcohol, 30 mL of water and a stir bar into the flask. Shake the flask gently to mix the reactants thoroughly. Place the flask in the cavity[49] of the microwave apparatus and attach two condensers to the flask.

2. Experimental Procedures

(1) Set the reaction temperature at 105 ℃, the power at a maximum of 300 W and the reaction time for 20 min. Turn on the microwave apparatus to complete the reaction.

(2) Take the flask out when the reaction is completed and filter the mixture under vacuum while it is hot.[50]

(3) If the filtrate is purple, add some saturated sodium bisulfite solution[51] until the purple color disappears.

(4) Transfer the colorless filtrate into a beaker and acidify the solution with 6 mL of concentrated hydrochloric acid until the solution is acidic (pH=3).[52]

(5) Collect the solid by vacuum filtration and wash it with some water and then air-dried.

(6) Weigh the product and calculate the percentage yield of benzoic acid.

3. Helpful Hints

(1) The microwave oven should be used properly and carefully following the teacher's instructions to avoid the explosions.[53] Be careful while removing the reaction flask from the oven as it may be extremely hot.

---

[48] 本实验采用：①加入过量的乙醇；②利用环己烷-水共沸脱水的方法，及时除去酯化过程中生成的水，来促使平衡反应右移，提高酯的收率。

[49] 腔体、箱内。

[50] 趁热过滤。

[51] 饱和亚硫酸氢钠溶液。

[52] 酸化的目的是将溶解在水中的苯甲酸的钾盐完全转化为苯甲酸，并以固体的形式析出。此步反应是放热的，盐酸应边搅拌边缓慢滴加。

[53] 微波炉应在老师指导下操作，严禁空载，以免发生爆炸。

(2) Make sure that there are not any cracks or chips in the glassware used for microwave-assisted reaction, which may shatter during the reaction.①

(3) Potassium permanganate is a strong oxidant. Avoid skin contact as it will stain your skin and clothing. Wash the affected area thoroughly with soap and water.

(4) Concentrated hydrochloric acid fume is suffocative and corrosive, so handle the HCl in the hood. If any acid spills on your skin, wash it off with large amount of water and then with dilute sodium bicarbonate solution.

(5) Add saturated sodium bisulfite solution to remove the excess potassium permanganate.

(6) The acidification is exothermic, so concentrated hydrochloric acid should be added slowly while stirring to avoid spraying.

(7) The crude benzoic acid can be purified by recrystallization from hot water.

- **Preparation of Ethyl Benzoate**

1. Setting Up

Place 5.0g of benzoic acid, 12.5 mL of ethanol, 3.0 mL of H_2SO_4, 10.0 mL of cyclohexane into a 100 mL three-necked flask, and mix the reactants thoroughly by gently swirling the flask.② Add several boiling chops into the flask and attach a reflux condenser to the flask. Heat the reaction mixture to boil slightly for 30 min. Allow the mixture to cool slightly, add several zeolites into the flask again, and then assemble the reaction apparatus as shown in Figure 4.10.

Figure 4.10 Apparatus for the preparation of ethyl benzoate

2. Experimental Procedures

(1) Add some water into the water separatory until the level is 0.5 mm lower than the top of its side neck.③

(2) Heat the reaction mixture slowly and keep refluxing gently for 1 h. Control the liquid level in the water separator carefully, making sure that the middle layer does not flow back to the flask, and the top layer of liquid is always a thin layer.④

(3) Stop heating. Allow the mixture to cool down and then pour the mixture into a beaker containing 40 mL of water.

(4) Transfer the mixture into a separate funnel. Shake the funnel well and stand it for a moment until the content in the funnel is separated into two distinct layers.

(5) Save the organic layer and extract the aqueous layer with diethyl ether (10 mL×2).

---

① 反应开始前要仔细检查玻璃仪器，确保所用的玻璃仪器没有破损或裂痕，以免在微波加热中发生炸裂。
② 反应物要混合均匀，避免局部浓度过高，受热炭化。
③ 分水器使用前需要检漏。
④ 随着酯化反应的进行，在分水器中会逐渐形成三层液体，上层和下层液体澄清透明，中层液体略显浑浊。反应过程要始终控制分水器中液面的位置，使最上层液体始终为薄薄的一层。

(6) Combine the ether extracts with the saved organic layer and wash the organic layer with about 20 mL of 10% sodium carbonate solution until the pH is 8-9.

(7) Wash the organic layer with 10 mL of water and transfer the resulting organic layer into a 50 mL clean and dry Erlenmeyer flask[⑧].

(8) Dry the solution over several amount of anhydrous sodium sulfate. Occasionally swirl the Erlenmeyer flask until the solution becomes clear.

(9) Decant the clear liquid into a dry round-bottom flask and assemble the flask for simple distillation to distill diethyl ether and the residual ethyl alcohol and cyclohexane. [⑨]

(10) Assemble the apparatus for vacuum distillation, distill the final product and collect the pure ethyl benzoate around 100 ℃ (20 mmHg)[⑩] in a pre-weighed receiving flask.

(11) Weigh the product and calculate the percentage yield of ethyl benzoate.

3. Helpful Hints

(1) Concentrated sulfuric acid is corrosive and can cause severe burns. Be extremely careful while handling it. If any concentrated sulfuric acid comes in contact with your skin, immediately wash it off with copious amounts of cold water[⑪] and then with dilute sodium bicarbonate solution.

(2) Mix the reaction mixture thoroughly and heat it to reflux gently to avoid the carbonization of organic compounds[⑫].

(3) As the reaction proceeds, the surface of the liquid in the water separatory will increase gradually and be separated into three layers. The upper layer contains a large amount of cyclohexane, some ethanol and a very small amount of water, while the middle layer has an increased relative content of water and a relatively reduced content of ethanol, and the lower layer is the original water in the water separatory. The top layer of liquid in the water separator is always a thin layer and the middle layer does not flow back to the flask, which ensures that most of the ethanol can overflow back into the reaction flask and the newly formed water stays in the water separator.

(4) Observe the level of the liquid in the water separatory carefully. Open the stopcock of the separatory and drip a few drops of water when the top layer of liquid in the water separator is higher than the side neck. [⑬]

(5) Some substances with a low boiling point should be distilled out by simple distillation before the finial vacuum distillation. Otherwise, it will cause damage to the oil pump.

- **Questions**

1. Can saturated solution of sodium thiosulfate be used to replace a saturated solution of

---

- ⑧ 锥形瓶。
- ⑨ 这一步操作的目的是将沸点低的有机物先通过简单蒸馏蒸出，以免在后续的减压蒸馏中对泵造成损害。
- ⑩ 减压蒸馏时，体系的实际压力有所不同，相应的沸点也不同。
- ⑪ 浓硫酸具有腐蚀性，如果不小心溅到皮肤上，应立即用大量冷水冲洗，然后再用稀的碳酸氢钠溶液冲洗。
- ⑫ 有机物炭化。
- ⑬ 反应过程中，要时刻关注分水器中液面的变化。当分水器中间层的液体即将要流回圆底烧瓶时，应打开分水器的活塞，放出少量的水，以保持最上层液体始终为薄薄的一层。

sodium bisulfite to make the purple color of the filtrate fade away in the preparation of the benzoic acid? Explain your answer.

2. Is it appropriate to add concentrated hydrochloric acid in the step of acidification until the solution becomes neutral(pH=7)? Explain the reason.

3. In the esterification reaction of benzoic acid and ethanol, the reaction mixture should be heated to reflux gently. What is the unexpected effect of flash heating?

4. There usually exists a reversible equilibrium in the esterification, and one of the methods to drive the equilibrium toward completion is to remove the newly formed product(s). What method is hereby employed to remove the by-product water? What is the purpose of adding cyclohexane to the reaction mixture?

5. What is the purpose of the following extracting and washing operation in the workup procedures for the esterification of benzoic acid to ethyl benzoate?

(1) Extracting the aqueous layer with ether(10 mL×2).

(2) Washing with 20 mL of 10% sodium carbonate.

(3) Washing the ether layer with 5 mL of water.

6. Discuss the differences observed in the IR spectra of benzoic acid and ethyl benzoate that are consistent with the formation of the latter in esterification of benzoic acid with ethanol.

Chapter 5

Experiments from Research and Literature

Exp.1 Synthesis of 3-Substituted Isoindolin-1-ones by Solvent-free Grinding Reaction[①]

- **Objectives**

1. To learn the method for preparation of 3-substituted isoindolin-1-ones via a cascade cyclization reaction.

2. To practice the solvent-free grinding reaction[⑮].

3. To practice the thin-layer chromatography technique.

- **Principle**

3-Substituted isoindolin-1-one represents one of the most important building blocks frequently found in bioactive compounds and natural products, and generally exhibits interesting pharmaceutical activity. Since the great value of this kind of molecule, significant efforts have been devoted by chemists to develop efficient methodologies for their synthesis. Over the past decades, contributions have been focused on solvent-required transformations as the vast majority, which are short of chemical economy due to the solvent-waste, laborious post-treatment (removal of solvent is generally required), and potential inflammable operations. The development of easily operated techniques for synthesis of valuable 3-substituted isoindolin-1-ones is therefore appealing.

In recent years, numerous outstanding works have been advanced by the utility of mechanochemical reactors[⑯]. These solvent-free mechanochemical reactions have been extensively applied in organic synthesis for preparation of structurally diverse organic compounds. As one of the techniques in mechanochemistry, grinding reaction is featured with solvent-free and easy operation. The target products could be readily obtained by grinding the starting materi-

---

[①] Reference: W P A, Yao L, Zhang D X, et al. Synthesis of 3-Substituted Isoindolin-1-ones by Solvent-Free Grinding Reaction [J]. University Chemistry, 2021, 36 (6), 2008040.

[⑮] 无溶剂研磨反应。

[⑯] 机械化学反应器。

als and some necessary reagents followed by the ease of post-treatment of the reaction mixture. In these grinding reaction events, a mortar[17] or ball mill[18] is utilized as the reaction container by a manual or automatic operation.

• **The Main Reaction**

$$\text{o-NC-C}_6\text{H}_4\text{-CHO} + \text{MeO}_2\text{C-CH}_2\text{-CO}_2\text{Me} \xrightarrow[\text{Grinding reaction}]{\text{K}_2\text{CO}_3, \text{Solvent-free}} \text{isoindolinone with } \text{CH(CO}_2\text{Me})_2$$

• **Apparatus**

Apparatuses required include mortar, beaker, glass rod, thin-layer chromatographic plates[19], capillaries[20], flash chromatographic column[21], 200-300 mesh silica gel, UV light.

• **Experimental Procedures**

1. Add 140 mg of anhydrous potassium carbonate(K_2CO_3) into a mortar with 12 cm diameter and grind quickly to produce a fine powder.[22]

2. Then weigh 1.31 g of 2-formylbenzonitrile and 1.39 g of dimethyl malonate, and transfer them into the mortar. Continue to grind the mixture at room temperature.

3. One minute later, the liquid substances are transformed into crude cake[23]. The powder is then obtained after grinding for 2 min more.

4. Monitor the reaction with a thin-layer chromatographic (TLC) plate developed by an eluting solvent of petroleum ether/ethyl acetate (2∶1). The product could be observed as a yellow pot on the TLC plate with retention factor value of 0.2.[24]

5. When 2-formylbenzonitrile is completely consumed, transfer the solid crude into a beaker and wash the remaining solid with 20 mL of dichloromethane twice (10 mL for each time).[25]

6. Add extra 40 mL of dichloromethane into the beaker to make sure the crude product is well-dissolved.

7. Assemble the flash chromatography apparatus and fill 1.0 cm-height 200-300 mesh silica gel[26]. Pour the solution into the prepared flash chromatographic column and pale-yellow filtrate is obtained.

8. After washing the silica gel with 10 mL of dichloromethane, the combined filtrate is

---

[17] 研钵。
[18] 球磨机。
[19] 薄层色谱板。
[20] 毛细管。
[21] 色谱柱。
[22] 研磨至细粉状。
[23] 粗产物为块状物。
[24] 产品是黄色固体，在薄层色谱板上可视。
[25] 通过紫外灯照射可判断邻甲酰基苯腈是否反应完全。
[26] 该步骤中1cm厚的硅胶层可起到滤纸的作用，用于过滤无机盐。

concentrated by rotary evaporator to give the 3-substituted isoindolin-1-one.

- **Helpful Hints**

1. Anhydrous potassium carbonate should be grinded into powder before adding the starting materials of 2-formylbenzonitrile and dimethyl malonate. The powder K_2CO_3 could promote the reaction.

2. The liquid starting materials will disappear along with the formation of the crude cake during the grinding. In this case, the reaction is not completed unless a further 2-minute grinding. The reaction is completed when a powder is formed.

3. 2-Formylbenzonitrile is colorless, the TLC plate should be put under the UV light to determine if 2-formylbenzonitrile is completely consumed.

4. The operation of pouring the solution into the flash chromatographic should be carried out more carefully to prevent breaking the silica gel.

- **Questions**

1. What is the alternative method for purifying the solid product with some impurities instead of chromatography? Point out the key operations for it.

2. What is the purpose of doing flash chromatography?

3. How to determine the purity of a solid product? Write down the corresponding techniques as many as you can.

Exp.2　Electroorganic Synthesis of Nitriles[1]

- **Objectives**

1. To know the principle for synthesis of nitrile by using benzaldehydoxime as the starting material.

2. To learn electroorganic chemistry[2] for organic synthesis.

- **Principle**

Compared with traditional organic synthesis, the electroorganic method represents a greener way to synthesize organic compounds by employing electricity as the renewable resource. Since no reagent waste is produced, the reaction features mild conditions, environment-benign, and easy manipulation. In 1834, Faraday first invented an electroorganic synthesis of ethane by electrolyzing an acetate solution[3]. Soon after in 1847, Kolbe developed the first useful transformation by converting salts of fatty acids into hydrocarbons via anodic oxidation. Over the years, electroorganic synthesis has become an attractive method for achieving various synthetic transformations including carbon-carbon bond formations, which received significant interest from academy and industry. With a basic understanding of elec-

---

[1] Reference: Marius F H, Siegfried R W. Electroorganic Synthesis of Nitriles via a Halogen-free Domino Oxidation-Reduction Sequence [J]. Chemical Communications. 2015, 51 (91): 16346-16348.

[2] 有机电化学。

[3] 电解乙酸溶液。

trochemistry, many reaction setups can be simplified using a simple battery as a power supply and two electrodes are then inserted into a solution that contains an electrolyte[50]. At one electrode(the anode) molecules are oxidized by an electron transfer from the reaction mixture to the electrode[51] while another electrode (the cathode) makes molecules reduced through an electron transfer from the electrode to the reaction mixture[52]. The net result is the transfer of electrons from the cathode to the anode.

Benzonitrile is one of the most important raw materials applied in the fields of pharmaceuticals, pesticides, and dyes[53]. In addition, it could also be utilized as useful synthetic intermediates capable of various transformations leading to corresponding carboxylic acid, amines, and amides. Although great advances have been achieved, reaction conditions using additional reagents such as oxidants or transition metals limit the application of these methods. Electroorganic synthesis is therefore an alternative way for synthesis of benzonitrile. With benzaldehydoxime as the substrate, benzonitrile could be yielded by a direct electroogranic process in undivided cells[54].

- **The Main Reaction**

- **Apparatus**

Apparatuses required include graphite anode, lead cathode, current density[55] of 10 mA/cm^2, an undivided 25 mL-beaker type cell, thin-layer chromatographic plates, capillary, flash chromatographic column, 200-300 mesh silica gel, UV light.

- **Experimental Procedures**

1. A solution of 2,4,6-trimethylbenzaldoxime(1.029 g, 6.305 mmol) and MTES(methyltriethyl ammonium methylsulfate, 0.576 g, 2.534 mmol)[56] in acetonitrile(MeCN, 25 mL) is electrolyzed at 22 ℃ with a current density of 10 mA/cm^2.

2. After the application of 2.5 F of charge, an excess of p-toluenesulfonyl hydrazide(0.128 g, 0.687 mmol) is added until no aldehyde appeared anymore on TLC detected by staining with 2,4-dinitrophenylhydrazine[57].

- [50] 电解质。
- [51] 在阳极，分子通过将电子转移至电极上而被氧化。
- [52] 在阴极，分子通过接收电极上的电子而被还原。
- [53] 药物、杀虫剂和染料。
- [54] 无隔膜电解槽。注：电解槽一般分为无隔膜电解槽和隔膜电解槽。
- [55] 电流密度。
- [56] 为该反应所用的电解质。
- [57] 加入对甲苯磺酰肼的目的是除去反应过程中肟水解的副产物醛。

3. When the reaction is completed, the solution is evaporated and the crude product is purified by column chromatography(cyclohexane/EtOAc 20∶1)to yield a white crystalline solid(m. p. 52 ℃)as product(0.740 g, 5.096 mmol, 81%).

- **Helpful Hints**

1. At the anode position, benzaldehydoxime is converted to nitrile oxide by the loss of two electrons, then at the cathode position, nitrile oxide is transformed into benzonitrile.

2. Be careful with the operation of the electricity equipment.

- **Questions**

1. Benzonitrile is an important organic chemical extensively applied in various areas, and many methodologies have been developed. Please search for traditional methods to synthesize benzonitrile in the literature.

2. Please consider the electron-transfer processes in the reaction from benzaldehydoxime to benzonitrile.

Exp.3 Ultrasonic Assisted Green Protocol for the Synthesis of N,N′-bis Phenylsulfamide[38]

- **Objectives**

1. To learn the principle for synthesis of sulfamide from sulfur chloride and aniline.
2. To learn the ultrasonic irradiation technique for organic synthesis.

- **Principle**

Ultrasonic irradiation[39] is a promising technique applied to efficient synthesis of various organic chemicals. Numerous successful examples indicate that the presence of ultrasound is able to enhance the rate of reactions, reduce the reaction time, and influence selectivity. These ultrasound irradiated transformations have been reported to occur with less expensive and less active reagents under mild conditions. Although this technique has been widely applied for a long period, the mechanism to enhance the reaction rate is still not very clear. Proposed mechanisms include an increase of mass-transfer coefficients caused by increased contact surface areas[40], stresses created after the implosion of cavities[41], and high temperature and pressure reached by cavitational collapses[42].

Sulfamides serve as significant functional groups frequently found in pharmaceuticals, pesticides, medicines, and drug candidates. The condensation between sulfonyl chloride and amine is one of the most utilized methods for their synthesis. In traditional conditions, the use of stoichiometric reagents such as base is necessary to promote the reaction yield, and the

---

[38] Reference: Ismahene G, Billel B, Khaoula B, et al. Ultrasonic Assisted Green Protocol for the Synthesis of Sulfamides [J]. Phosphorus, Sulfur, and Silicon and the Related Elements. 2022, 197 (11): 1150-1156.

[39] 超声波照射。

[40] 通过增加接触表面积提高传质系数。

[41] 空腔内爆后产生的应力。

[42] 空化塌陷产生高温高压。

reaction time is relatively long. In total, reactions in a traditional manner exhibit low efficient to produce sulfamides in harsh conditions and low conversion. In this reaction, sulfuryl chloride(SO_2Cl_2) reacts efficiently with aniline by ultrasonic irradiation to afford N,N'-bis phenylsulfamide.

- **The Main Reaction**

$$2\ PhNH_2 + SO_2Cl_2 \xrightarrow[RT]{45\ kHz} Ph-NH-SO_2-NH-Ph$$

- **Apparatus**

Apparatuses required include ultrasonic bath with a frequency of 45 kHz and an output power of 250 W, 10 mL round-bottom flask, Büchner funnel and flask.

- **Experimental Procedures**

1. Add aniline(2 mmol) and sulfuryl chloride(1 mmol) in a 10 mL round-bottom flask.

2. Then the reaction mixture is subjected to ultrasonication by an ultrasonic bath at a frequency of 45 kHz at 40℃ for 40 min[33].

3. When the reaction is completed, the mixture is crystallized in diethyl ether[34] and the product is obtained by filtration.

- **Helpful Hints**

1. Sulfuryl chloride(SO_2Cl_2) is a toxic liquid bearing irritating odor[35], measure and use this reagent in the fume hood carefully.

2. Since both of two starting materials are liquids, the end of the reaction is determined by the consumption of the liquids in the flask.

- **Questions**

1. Summarize the literature methods for synthesis of sulfamides.

2. How to further purify the sulfamide product by recrystallization? Design the apparatus and point out main operations.

Exp.4 Clean Synthesis in Water: Uncatalysed Preparation of Ylidenemalononitriles[36]

- **Objectives**

1. To know the principle for synthesis of benzylidenemalononitriles via a condensation reaction.

---

[33] 超声反应过程中，需要适当振荡或搅拌。

[34] 先将产物溶解在乙醚溶剂中，过滤后，待乙醚缓慢挥发即可得到晶状产物。

[35] 磺酰氯具有刺激性气味，容易水解。

[36] Reference: Franca B, Maria L C, Raimondo M, et al. Clean Synthesis in Water: Uncatalysed Preparation of Ylidenemalononitriles [J]. Green Chemistry. 2000, 2 (3): 101-103.

2. To learn the operation of organic reaction in greener water solvent.

• **Principle**

To meet the requirements of developing green organic chemistry, great efforts have been devoted by chemists to inventing synthetic methodologies for synthesis of value-added organic compounds in high efficiency, high atom-economy, and environmentally benign. Thus, the attempt to reduce the usage of expensive reagents, harsh conditions, difficult operations becomes the main goal of organic synthesis in academy and industry. Currently, homogenous organic reactions are generally performed in organic solvents due to the good solubility of raw materials. This therefore leads to big problems of wasting a large amount of organic solvents and causing additional environmental pollution during the solvent disposal. Water is a green reagent and ubiquitous all over the world, therefore, water as the reaction media is appealing in organic chemistry.

In recent years, more and more synthetic reactions have been reported to be well achieved in water solvent. Upon the completion of the reaction, the product would be obtained only by a simple extraction or filtering operation[47]. Benzylidenemalononitrile was reported to be effectively applied in anti-fouling agents, fungicides and insecticides[48]. The condensation of benzaldehyde and malononitrile represents the direct route to prepare this molecule. Precedent reports were realized in organic solvents generally by the promotion of bases such as piperidine, sodium ethoxide, or sodium hydroxide. Besides, Lewis acids were also reported to enable this transformation. In these cases, stoichiometric reagents and a lot of organic solvents are inevitably utilized. As a cleaner methodology for synthesis of benzylidenemalononitrile, the reaction in greener water solvent has been achieved by heating operation.

• **The Main Reaction**

$$\text{PhCHO} + \text{NC-CH}_2\text{-CN} \xrightarrow[65\ ^\circ\text{C},\ 1\ \text{h}]{\text{H}_2\text{O}} \text{PhCH=C(CN)}_2 + \text{H}_2\text{O}$$

• **Apparatus**

Apparatuses required include 50 mL round-bottom flask, Büchner funnel and flask.

• **Experimental Procedures**

1. Add benzaldehyde(10 mmol), malononitrile(0.66 g, 10 mmol), and 10 ml of distilled water[49] in a round-bottom flask.

2. The reaction is then heated to 65 ℃ for 1 h. Upon the completion of the reaction, the reaction system is cooled to 10 ℃[50].

3. Benzylidenemalononitrile product is then obtained by Büchner filtration.

---

[47] 通过简单的萃取或过滤操作即可得到产物。
[48] 防污剂、杀菌剂和杀虫剂。
[49] 蒸馏水可通过常压蒸馏获得。
[50] 将反应混合物冷却至10℃,以确保产物能充分析出。

● **Helpful Hints**

1. Distilled water should be prepared by simple distillation before the reaction to make sure of a good efficiency.

2. The crude product should be dried under infrared red light before calculating the percentage yield.

● **Questions**

1. Benzylidenemalononitrile is a solid compound with a minor solubility in cold water, how to determine the purity of the product and how to further purify it by using organic chemistry techniques?

2. How to determine the end of this condensation reaction for synthesis of benzylidenemalononitrile?

Exp.5 Transition-metal-catalyzed Convenient Synthesis of Internal Alkyne[⑮]

● **Objectives**

1. To know the method for synthesis of internal alkyne.

2. To learn the transition-metal-catalyzed cross-coupling reaction[⑯].

3. To understand the principle of Pd/Cu-cocatalyzed[⑰] Sonogashira reaction.

4. To learn the operation of oxygen sensitive reaction with Schlenk line technique[⑱].

● **Principle**

Alkyne is an important unsaturated organic compound bearing a carbon-carbon triple bond. Transformation of the triple bond of alkynes enables facile synthesis of structurally diverse organic compounds such as substituted alkenes and substituted alkanes by the functionalization of π-bonds, and diversified cyclic compounds by cycloaddition reactions of π-bonds with 1, n-dipolar. For examples: (1) alkynes could be converted into alkenes in the presence of Lindlar catalyst; (2) the copper-catalyzed cycloaddition reaction of alkynes with azides would yield heterocyclic 1,2,3-triazoles, which is known as a Click chemistry awarded as Nobel prize in 2022[⑲]. Therefore, the synthesis of alkynes is of great importance in organic synthesis.

Transition-metal-catalyzed cross-coupling reaction represents the most significant way to form a carbon-carbon bond. In this regard, Richard Heck, Ei-ichi Negishi and Akiro Suzuki

---

⑮ Reference: Kenkichi S, Yasuo T, Nobue H. A Convenient Synthesis of Acetylenes: Catalytic Substitutions of Acetylenic Hydrogen with Bromoalkenes, Iodoarenes and Bromopyridines [J]. Tetrahedron Letters. 1975, 16 (50): 4467-4470.

⑯ 过渡金属催化的交叉偶联反应。

⑰ 钯/铜-共催化；反应中，钯催化剂和铜催化剂分别起不同作用，共同实现反应的高效转化。

⑱ "希莱克技术"（双排管操作技术），广泛应用于无水无氧反应中。

⑲ 点击化学，是由化学家卡尔·巴里·夏普莱斯（K. B. Sharpless）在 2001 年提出的一个合成概念，在 2022 年获得诺贝尔化学奖（这也是 Sharpless 第二次获得诺贝尔奖）。

were rewarded as Nobel prize in 2010 due to their great contributions to Pd-catalyzed cross-coupling reactions in organic synthesis. As one of the important cross-coupling reactions, the Sonogashira reaction❶ has also been well-known attributed to its capacity for synthesis of alkynes since the seminar work reported by a Japanese organic chemist Kenkichi Sonogashira in 1975. In general, internal alkynes can be synthesized from the coupling of terminal alkynes with aryl or alkenyl halides by forming a carbon-carbon bond under the Pd/Cu-cocatalyst. Over the past decades, Sonogashira reaction has been applied in many fields including total synthesis of natural products, drug discovery, and preparation of functional materials. In recent years, an improvement of this reaction has also been achieved in exploring greener reaction conditions, which would make the construction of internal alkynes much more convenient. ❷

• **The Main Reaction**

$$\text{PhI} + \text{PhC}\equiv\text{CH} \xrightarrow[\text{Et}_2\text{NH}]{\text{PdCl}_2(\text{PPh}_3)_2, \text{CuI}} \text{Ph}-\text{C}\equiv\text{C}-\text{Ph}$$

• **Apparatus**

Apparatuses required include Schlenk line, nitrogen cylinder, Schlenk tube, oil pump, short silica gel column.

• **Experimental Procedures**

1. Add $\text{Pd}(\text{PPh}_3)_2\text{Cl}_2$ (0.005 mmol) and CuI (0.01 mmol) to a Schlenk tube with a stirring bar.

2. Connect the Schlenk tube with the Schlenk line attached to nitrogen cylinders and oil pump. Check and make sure all joints are well-connected, stopcock of nitrogen cylinder is closed, and the Schlenk tube is half-open❸.

3. Open the oil pump and keep the connection of Schlenk tube with oil pump to make a vacuum system in Schlenk tube. ❹

4. Then turn the connection of Schlenk tube to nitrogen cylinder and open the tank valve of nitrogen cylinder and the regulator needle valve to adjust the stream of nitrogen gas from the cylinder. ❺

5. When the Schlenk tube is fully-purged with nitrogen, close the regulator needle valve and repeat the above operations three times. ❻

6. When these operations are finished, keep purging nitrogen and open the Schlenk tube

---

❶ Sonogashira 反应在取代炔烃以及大共轭炔烃的合成中得到了广泛的应用,并在很多天然化合物、农药医药、新兴材料等分子的合成中起着关键的作用。

❷ 近年来,Sonogashira 偶联反应得到进一步发展,反应条件更为绿色。

❸ Schlenk tube(史莱克管)是指带有支口的可密封反应管,可用作双排管技术的反应容器;当史莱克管处于半开状态时,反应管口关闭,支口打开。

❹ 该操作将史莱克管与油泵连通,抽去史莱克管内空气。

❺ 该操作将史莱克管与氮气钢瓶连通,使史莱克管内充满氮气。

❻ 史莱克管中气体一般需要重复置换三次,以确保管中的空气完全被氮气置换。

immediately.① Add phenylacetylene(1 mmol), iodobenzene(1 mmol), and triethylamine(10 mL) into the Schlenk tube. Close the Schlenk tube and stopcocks of nitrogen cylinder at the same time②.

7. The reaction mixture is then allowed to stir at room temperature for 6 h. When the reaction is completed, remove triethylamine under reduced pressure and add cyclohexene to dissolve the solid.

8. By passing over a short silica gel column to remove the catalyst③, the crude solution is obtained, which should be concentrated under reduced pressure and further purified by crystallization with alcohol to give the product diphenylacetylene.

- **Helpful Hints**

1. Be careful with the Schlenk line operation, especially when adjusting the regulator needle valve to keep a suitable stream of nitrogen.

2. The reaction is sensitive to oxygen due to the side reaction of Glaser reaction occurred in the presence of oxygen④.

3. Except for the $PdCl_2(PPh_3)_2$ catalyst, other palladium catalysts such as $Pd(PPh_3)_4$ are also suitable for this cross-coupling reaction.

- **Questions**

1. Why does the exchange of air in the Schlenk tube with nitrogen by Schlenk line system require three-times operation?

2. What is the result if the operation of passing over a short silica gel column is not conducted?

3. How to determine the end of this Sonogashira reaction. Write down the operation details.

Exp.6 N-Heterocyclic Carbene Catalyzed Intramolecular Benzoin Condensation to Access 9,10-Phenanthraquinone⑤

- **Objectives**

1. To learn the method for synthesis of 9,10-phenanthraquinone.
2. To know about the organocatalyst⑥ of N-heterocyclic carbene.
3. To learn an intramolecular benzoin condensation reaction.

- **Principle**

Benzoin condensation represents one of the most important C—C coupling reactions. By using two molecules of aldehydes as starting materials, structurally significant α-hydroxyketones could be achieved with great efficiency. Benzoin condensation was first discovered in

---

① 打开氮气钢瓶减压阀后,应立即打开史莱克管,以避免管内压力过大,产生危险。
② 关闭史莱克管时,应立即关闭氮气钢瓶减压阀,以避免管内压力过大,产生危险。
③ 催化剂可通过一个短的硅胶柱除去。
④ 为了避免Glaser副反应的发生,Sonogashira需要在惰性气体氛围下进行。
⑤ Reference: Dieter E, Oliver N. Thiazol-2-ylidene Catalysis in Intramolecular Crossed Aldehyde-ketone Benzoin Reactions Synlett, 2004, 12: 2111-2114. (氮杂环卡宾催化的分子内安息香缩合反应制备9,10-菲醌)
⑥ 有机催化剂。

1832 by Liebig and Wöhler, and the first methods were only suitable for the conversion of aromatic aldehydes. The mechanism of this well-known cyanide-catalyzed condensation⁸ was proposed by Lapworth in 1903. With cyanide ion as the catalyst, conversion of benzaldehyde leads to α-hydroxy benzoin through a cyanohydrin intermediate⁹ followed by a tetrahedral nucleophilic acyl anion⁷⁰ after deprotonation, thus exhibiting a reversal of the original electrophilic carbonyl group of benzaldehyde known as umpolung⁷¹. Undoubtedly, the use of strongly toxic cyanide ions as the catalyst is the terminal limit for this transformation.

Fortunately, organocatalysis serves as an attractive alternative in this benzoin condensation reaction. A particular hallmark is the usage of *N*-heterocyclic carbene (NHC) as the catalyst. In 1943, Ukai and coworkers demonstrated that the NHC precursor thiazolium salts⁷² could be used as catalysts in the benzoin condensation reaction. The mechanism of this process was investigated by Breslow, who envisaged the formation of a nucleophilic *N*-heterocyclic carbene (NHC) from azolium salt under basic conditions and proposed a Breslow intermediate⁷³. The nucleophilic attack of the Breslow intermediate to another molecule of aldehyde furnished the benzoin via a umpolung of aldehydes. Generally, imidazolium, thiazolium, or triazolium salt-derived NHCs have been used successfully for this umpolung reactions. Since benzoin condensation creates a chiral center on the carbon bearing the —OH group, the use of chiral NHC catalysts allowed the enantioselective variants of benzoin condensation.

9,10-Phenanthraquinone is an important chemical in organic synthesis, which is capable of synthesis of benthanthrone⁷⁴. 9,10-Phenanthraquinone bears bacteriostatic ability, and can be used in seed dressing to prevent grain smut and cotton seedling disease. From the viewpoint of organic synthesis, it can be achieved through the oxidation of its corresponding α-hydroxy benzoin with the oxidants such as nitric acid, $FeCl_3$, Fe_2O_3, or $ZnNO_2$-SiO_2. In this experiment, 9,10-phenanthraquinone is afforded from the corresponding α-hydroxy ketone in situ generated by NHC-catalyzed benzoin condensation of [1,1′-biphenyl]-2,2′-dicarbaldehyde with oxygen as the oxidant. The thiazolium salt is used as the precatalyst.

- **The Main Reaction**

---

⁶⁸ 传统的安息香缩合是指芳醛（不含 α-H），在含水乙醇中以氰化钠或氰化钾为催化剂，加热后发生自身缩合，生成 α-羟基酮的反应。

⁶⁹ 氰醇负离子中间体。

⁷⁰ 在氰醇负离子中，连有氰基的碳原子上的氢酸性很强，在碱性介质中会转化形成亲核性的氰醇碳负离子。

⁷¹ 在氰醇碳负离子中，原本呈正电性的羰基碳原子变为负电性，表现出亲核性，这一变化称为极性反转。

⁷² 噻唑盐。

⁷³ 布雷斯洛中间体。

⁷⁴ 苯绕蒽酮。

• **Apparatus**

Apparatuses required include round-bottom flask, nitrogen cylinder, reflux condenser, oil pump, chromatography column.

• **Experimental Procedures**

1. Place 20 mmol % NHC catalyst precursor, t-BuOH (0.1 mol/L), and [1,1'-biphenyl]-2,2'-dicarbaldehyde under nitrogen atmosphere in a round-bottom flask with a reflux condenser.

2. Heat at 60℃ after the addition of 30 mol % DBU base to the solution stirred at this temperature.

3. Monitor the reaction by TLC until the full consumption of the starting material.

4. After the completion of the reaction, the mixture was then poured into H_2O, extracted twice with CH_2Cl_2, dried over $MgSO_4$, and concentrated under reduced pressure.

5. The residue was purified by flash chromatography (silica gel, Et_2O/pentane).

• **Helpful Hints**

1. Be careful with the Schlenk line operation, especially when adjusting the regulator needle valve to keep a suitable stream of nitrogen.

2. The free NHC catalyst should be in situ generated by the treatment of the catalyst precursor with base under the nitrogen atmosphere due to its sensitive to air.

• **Questions**

1. Why can the α-hydroxy benzoin in this experiment be easily oxidized to 9,10-phenanthraquinone in the air?

2. How does the catalyst precursor thiazolium salt transfer into the active NHC catalyst in the presence of base.

3. Try to provide a rapid and efficient method to synthesize the starting material [1,1'-biphenyl]-2,2'-dicarbaldehyde.

Exp.7　Optical Resolution of (±)-(S)-Methyl-(S)-phenylsulfoximine with (+)-L-Camphorsulfonic Acid[①]

• **Objectives**

1. To learn the method for synthesis of chiral (S)-methyl-(S)-phenylsulfoximine.

2. To learn the optical resolution technique for separation of enantiomers.

3. To preview the separation techniques of chromatography.

• **Principle**

Chirality is a universal phenomenon in nature and has extensive application in many

---

[①] Reference: Jochen B, Hans-Joachim G An Efficient Resolution of (±)-(S)-Methyl-(S)-phenylsulfoximine with (+)-L-Camphorsulfonic Acid by the Method of Half-quantities[J]. Tetrahedron: Asymmetry. 1997, 8(6): 909-912. [以 (+)-L-樟脑磺酸为拆分试剂对(±)-(S)-甲基-(S)-苯基亚砜亚胺进行手性拆分]

fields. For instances, natural materials including minerals, organic molecules, and biological structures may possess unique chirality. The development of methods to access chiral compounds represents one of the most highly urgent and desirable missions in organic synthesis. Although the synthesis of racemic compounds is relatively simple, additional techniques are needed to give optically active compounds. Optical resolution is one of the significant techniques, and the first resolution was carried out by Louis Pasteur[36] in 1848, when he digested racemic tartaric acid with the mold Penicillium glaucum and observed the unnatural (S),(S) enantiomer remained unchanged. Soon later in 1899, Pope and Peachey described the "half equivalent" method, which is one of the most economical methods of resolution. As a separation method, optical resolution is a more cost-effective approach on a large scale than enantioselective synthesis and chromatographic separation methods popularly developed in modern chemistry.

The essence of resolution is the differential interaction of the resulting agents with the couple of enantiomers in a racemic mixture. A pair of diastereomers formed could be separated by achiral methods[37]. Finally, decomposition of the pure diastereomers occurs to afford the pure enantiomers and release the resolving agent. During this process, several problems should be solved:(1) what kind of resolving agent is suitable for the resolution;(2) what's the molar ratio of racemate and resolving agent;(3) which is the optimal solvent;(4) how to separate the formed diastereomers.

In this experiment, racemic (±)-(S)-methyl-(S)-phenylsulfoximine is treated with (+)-camphorsulfonic acid((+)-CSA) to result in the separation of the two enantiomers, leading to(+)-(S)-methyl-(S)-phenylsulfoximine and(−)-(S)-methyl-(S)-phenylsulfoximine respectively.

- **The Main Reaction**

- **Apparatus**

Apparatuses required include round-bottom flask, Büchner funnel, filter flask, acidic cation exchanger(Lewatit S100).

- **Experimental Procedures**

1. A solution of (+)-camphorsulfonic acid((+)-CSA)(87.3 g, 376 mmol) in dry

---

[36] 路易·巴斯德（1822—1895）是法国微生物学家、化学家，美国学者麦克·哈特所著的《影响人类历史进程的100名人排行榜》中，巴斯德名列第12位，其发明的巴氏消毒法直至现在仍在应用。

[37] 手性拆分的关键是选择合适的具有手性的拆分试剂，并让其与一对对映异构体发生反应，生成一对非对映异构体。非对映异构体具有不同的物理性质和化学性质，可通过非手性的方法分离。

acetone(550 mL) is added gradually at room temperature under stirring to a solution of racemic(±)-(S)-methyl-(S)-phenylsulfoximine (116.6 g, 751 mmol) in dry acetone (430 mL).

2. After the addition of about one-third of the solution of (+)-CSA, fine white crystals of the salt (+)-sulfoximine/(+)-CSA begin to precipitate.

3. The resulting suspension is stirred at room temperature for 12 h.

4. The crystals are filtered with the aid of a glass filter, washed thoroughly with dry acetone(4×100 mL) and dried in vacuum to give the salt (+)-sulfoximine/(+)-CSA (123.0 g, 84%).

5. Base treatment of the salt (+)-sulfoximine/(+)-CSA and distillation(89 ℃, 0.1 Torr) gives the (+)-sulfoximine(47.3 g, 96%).

6. The filtrate remaining from the above isolation of (+)-sulfoximine/(+)-CSA, which contains (−)-sulfoximine(79% ee), is treated with solid(+)-CSA(17.5 g, 75.4 mmol).

7. After the concentration of the solution in vacuum(40 ℃, Torr)[75], the resulting oil is treated under stirring with dry acetone(2 mL) and dry toluene(800 mL) whereby a fine suspension is formed.

8. The suspension is stirred at room temperature for 1 d and kept at 2 ℃ for 1 d.

9. Filtration and drying of the solid material in vacuum give the salt (−)-sulfoximine/(+)-CSA(5.6 g).

10. Concentration of the filtrate and distillation gives the (−)-sulfoximine(43 g, 74%).

11. The basic aqueous solution, which remained from the base treatment of the salt (+)-sulfoximine/(+)-CSA and extraction of the (+)-sulfoximine, is passed through a column containing an acidic cation exchanger[76](Lewatit S100).

12. Concentration of the obtained solution in vacuum and recrystallization of the residue from ethyl acetate give(+)-CSA in 97% yield.

• **Helpful Hints**

1. 0.5 equivalent of(+)-camphorsulfonic acid should be used as the resulting agent to make sure of the completion of the optical resolution and give the(+)-sulfoximine in excellent ee value.

2. Acidic cation exchanger(Lewatit S100) is needed to achieve the recovery of(+)-camphorsulfonic acid.

3. The purpose of procedures 6-10 is to remove the remaining(+)-sulfoximine in the filtrate, and to give the(−)-sulfoximine with excellent ee value.

• **Questions**

1. What if more than 0.5 equivalent of(+)-camphorsulfonic acid is used as the resulting agent?

---

[75] 通过减压蒸馏分离。
[76] 酸性阳离子交换剂。

2. Except the vacuum distillation presented in this experiment, whether other methods can be used to separate the enantiomers of (+)-sulfoximine and (−)-sulfoximine, try to list the possible methods.

3. Explain the principles of the optical resolution.

Exp.8 Synthesis of Benzyl Acetate in Room Temperature Ionic Liquids[①]

- **Objectives**

1. To understand the definition and characteristics of ionic liquids[②].
2. To learn the principle and method of preparing ionic liquids.
3. To learn the method of preparing an ester by the reaction of an alkyl halide with a carboxylate using an ionic liquid as solvent and catalyst.
4. To practice the washing and extraction of liquid compounds.
5. To learn the qualitative and quantitative analysis[③] of organic compounds using chromatographic methods.

- **Principle**

The most common definition of ionic liquids is refers to the salts that exist in liquid state below 100 ℃. In fact, the most used ionic liquids are liquids at room temperature. Compared with traditional volatile organic solvents, the ionic liquids at room temperature are new types of solvents and are endowed with some advantages, such as near-zero vapor pressures[③], environmentally benign, wide range of polarity and recyclability[④]. Ionic liquids are also called "task-specific" or "designer" solvents because both the cation and anion in an ionic liquid can be altered relatively easily, thereby causing changes in the properties of the ionic liquids[⑤]. A variety of organic, organometallic and polymeric compounds are well solvated in ionic liquids, which makes ionic liquids widely used as solvents and catalysts in organic synthesis.

The imidazolium-based ionic liquids[⑥] are widely used in organic synthesis, and are usually prepared by the following two steps: the first is the quaternization[⑦] of imidazole and an alkyl halide to form the corresponding quaternary ammonium salt of [CnMin][X]. Then, anion metathesis[⑧] is carried out to change the anion X^- to the desired one, such as BF_4^- and

---

① Reference: Jonathan G, Huddleston A E, Visser W, et al. Characterization and Comparison of Hydrophilic and Hydrophobic Room Temperature Ionic Liquids Incorporating the Imidazolium Cation [J]. Green Chemistry. 2001, 3: 156-164.

② 离子液体一般不会变为蒸气，其蒸气压接近于零。因而，离子液体在使用过程中不会产生对大气造成污染的有害气体。

③ 定量和定性分析。

④ 离子液体具有可设计性。通过改变阳离子和阴离子的不同组合，可以设计出各种种类的离子液体。

⑤ 极性范围广、可回收。

⑥ 离子液体是指全部由离子组成的液体。离子液体具有独特的性能，被广泛应用于化学研究的各个领域中。其中，离子液体作为传统有机溶剂的绿色替代品，已应用到多种类型的有机反应中。

⑦ 咪唑类离子液体。

⑧ 季铵化作用。

⑨ 阴离子交换。

CF_3COO^-, etc. For example, the ionic liquids, $[C_4Min][BF_4]$ is prepared as below.

Benzyl acetate is one of the most widely used fragrances, mainly prepared in the laboratory by the nucleophilic substitution reaction[①] of benzyl chloride and sodium acetate under basic conditions. However, this synthesis method involves the use of triethylamine, pyridine and other volatile and toxic substances, and the demands for high temperature and lengthy reaction time. In this experiment, the reaction of sodium acetate and benzyl chloride to form benzyl acetate is performed in the ionic liquids $[C_2Min][BF_4]$ (1-ethyl-3-methylimidazolium tetrafluoroborate), which is found to be an effective solvent and catalyst for this reaction.

- **The Main Reaction**

- **Apparatus**

Apparatuses required include 50 mL round-bottom flask, separatory funnel, pressure equalizing addition funnel, reflux condenser, cylinder, rotary evaporator and apparatus for magnetic stirring.

- **Experimental Procedures**

1. Preparation of the ionic liquid[C_4Min][BF_4]

(1) Place 10.5 mL of 1-chlorobutane and 8.7 mL of 1-methylimidazole[②] into a round-bottom flask.

(2) Add a stir bar into the flask and assemble the reflux apparatus.

(3) Heat the reaction mixture to 70 ℃ and react for 24-27 h while stirring until the two layers are formed in the mixture.

(4) Decant the upper layer slowly, and then add 15 mL of ethyl acetate to the lower layer and swirl the flask to mix thoroughly[③].

(5) Decant the upper layer of ethyl acetate slowly and add 15 mL of fresh ethyl acetate to the mixture again.

(6) Repeat the above procedures several times to remove the unreacted starting materials in the lower layer[④].

(7) Add 30 mL of water to the resulting mixture to precipitate the yellow crystal of $[C_4Min][Cl]$ after the final decanting.

(8) The ethyl acetate used in the above steps can be recycled and treated for further use.

---

① 亲核取代反应。
② 1-甲基咪唑。
③ 振摇反应瓶，使液体混合均匀。
④ 通过多次洗涤，可除去溶解在乙酸乙酯中未反应的原料。

(9) Transfer the obtained [C$_4$Min][Cl] into a plastic bottle (lined with perfluorinated materials)[93] and add 17.5 mL of 42% HBF$_4$ solution. Then, stir the mixture for 24 h.

(10) After the reaction is complete, transfer the mixture to a separatory funnel and extract it with dichloromethane(30 mL×3).

(11) Combine the dichloromethane extracts and dry the solution over anhydrous magnesium sulfate.

(12) Decant the dried dichloromethane solution into a dried round-bottom flask carefully. Assemble the flask for rotary evaporation to remove the solvent completely.

(13) The residue in the round-bottom flask is the ionic liquid [C$_4$Min][BF$_4$].

2. Preparation of benzyl acetate

(1) Place 5 mL of ionic liquid [C$_4$Min][BF$_4$], 4.2 mL of benzyl chloride and 3.5 g of sodium acetate into a round-bottom flask. Add a stir bar into the flask and assemble the reflux apparatus.

(2) Heat the reaction mixture in an oil-bath for 2 h while stirring, with the reaction temperature kept at 60 ℃.

(3) Cool down the mixture to room temperature after the reaction is completed.

(4) Extract the resulting mixture with toluene(10 mL ×3).

(5) Combine the toluene extracts and conduct the qualitative and quantitative analysis of benzyl acetate using chromatographic methods[94].

3. Reusing of the ionic liquid

(1) Extract the ionic liquid obtained after separation with toluene(10 mL×3) to remove some residual organic compounds.

(2) Maintain the resulting ionic liquid at 80 ℃ for 30 min under vacuum(20 mmHg) after separating the toluene.

(3) The obtained ionic liquid can be further used for the reaction directly.

• **Helpful Hints**

1. 1-methylimidazole and 1-chlorobutane are both toxic and can cause serious damage to the upper respiratory tract, skin and eyes. Wear gloves and protective glasses while handling them. Perform the reaction in the hood to minimize exposure to it.

2. Ethyl acetate, dichloromethane and toluene are all toxic and volatile organic solvents. Use these solvents in a well-ventilated area[95] and avoid contact or inhalation for a long period[96].

3. In the preparation of benzyl acetate, the reaction should be performed at 60 ℃. At this

---

[93] 内衬全氟材料的塑料瓶。

[94] 色谱法。

[95] 通风良好的区域。

[96] 避免长期接触或吸入。

reaction temperature, $[C_4MIm]BF_4$ can be miscible① with molten sodium acetate into a single phase. After the completion of the catalytic reaction, cool down the mixture to room temperature (below 25 ℃), and the formed sodium chloride will be solidified to the bottom of the reactor. Meanwhile, the product benzyl acetate is insoluble in ionic liquids and can be easily separated by extraction with toluene.

4. The anion metathesis reaction of $[C_4Min][Cl]$ with HBF_4 cannot be performed using glassware.

• **Questions**

1. What are the characteristics of the ionic liquids?

2. What is the significance of anion metathesis in the synthesis of ionic liquids?

3. Library project: Look up the traditional synthesized methods for the preparation of benzyl acetate by the reaction of benzyl chloride and sodium acetate. Try to summarize the advantages of the method used in this experiment compared to other traditional ones?

4. Do you think ionic liquids are green solvents? Why or why not?

Exp.9   Green Synthesis of 4-Bromoaniline②

• **Objectives**

1. To study the method of green synthesis of 4-bromoacetanilide by electrophilic aromatic substitution reaction.

2. To understand the substituent effects in electrophilic substitutions of aromatic rings.

3. To study to remove acetyl group under acidic conditions.

4. To practice refluxing and simple distillation.

5. To practice the vacuum filtration and washing of solid compounds.

• **Principle**

Electrophilic aromatic substitution is an important reaction of aromatic compounds, which allows the introduction of many different functional groups onto an aromatic ring. For example, bromobenzene can be obtained by the electrophilic substitution of bromine and benzene in the presence of Lewis acids. If there is an electron-donating group③on the aromatic ring, the bromination reaction will be performed easily under mild conditions. For instance, acetanilide can be converted to 4-bromoacetanilide using molecular bromine in acetic acid. Given that acetamido group (CH_3CONH—) activates the benzene ring toward electrophilic substitution, a Lewis acid catalyst is necessarily required. The acetamido group is an *ortho*- and *para*-directing moderate activator④, so selective monobromination of the acetanilide oc-

---

① 混溶的。

② Reference: Biswas R. A Mukherjee Introducing the Concept of Green Synthesis in the Undergraduate Laboratory: Two-Step Synthesis of 4-Bromoacetanilide from Aniline. [J] J. Chem. Educ. 2017, 94 (9): 1391-1394.

③ 供电子基团。

④ 中等强度的邻、对位致活基团。

curs and a mixture of 2-bromo- and 4-bromoacetanilide can be obtained. However, the steric bulk of the acetamido group hinders attack at the 2-position, so 4-bromoacetanilide is expected to predominate as the major product.

$$\text{acetanilide} \xrightarrow[\text{CH}_3\text{COOH}]{\text{Br}_2} \text{4-bromoacetanilide (Major)} + \text{2-bromoacetanilide (Minor)}$$

The greener replacement of reactants is one of the most active areas of research in green organic chemistry. Bromine is a hazardous chemical that may cause serious chemical burns. To avoid the hazards to human health and environment associated with the direct application of bromine, the bromine for this reaction is generated in situ by an oxidation-reduction reaction of 40% hydrobromic acid with hydrogen peroxide (H_2O_2). Hydrogen peroxide is a green oxidizing agent, and its reduction product is water, which makes the synthetic route more in line with the requirements of green chemistry[①]. The reaction is shown as below:

$$\text{acetanilide} \xrightarrow[\text{EtOH}]{\text{HBr}, H_2O_2} \text{4-bromoacetanilide}$$

Acyl group can be removed to generate the amino group by hydrolyzing[②] the amide. The amide bond is inert under acidic or basic conditions, so this conversion is accomplished under a harsh condition by refluxing with the strongly acidic or basic solution. In this experiment, deprotection of the acetyl group from 4-bromoacetanilide to form 4-bromo aniline is accomplished by refluxing with HCl. The corresponding reactions are shown as follows.

$$\text{4-bromoacetanilide} \xrightarrow[\Delta]{H^+} \text{4-bromoanilinium} + CH_3COOH$$

- **Apparatus**

Apparatuses required include 150 mL three-necked flask, Erlenmeyer flask, pressure equalizing addition funnel[③], reflux condenser, long-stem funnel[④], distillation head, condenser, thermometer, thermometer adapter[⑤], adapter, beaker, stopper, Büchner funnel, filter flask and apparatus for magnetic stirrer.

- **Experimental Procedures**

1.Preparation of 4-bromoacetanilide

(1) Weigh 10 g of acetanilide into a three-necked flask that contains a stir bar.

(2) Add 60 mL of 95% ethanol and swirl the mixture in the flask until all of the solids

---

① 符合绿色化学的发展需求。
② 在碱性或酸性条件下将酰基水解，可重新得到氨基。
③ 恒压滴液漏斗。
④ 长柄漏斗。
⑤ 温度计套管。

are dissolved[①].

(3) Add 17 g of 40% hydrobromic acid to the flask.

(4) Place 17 g of 30% hydrogen peroxide in the pressure equalizing addition funnel. Turn on the stirrer and add hydrogen peroxide dropwise to the stirred reaction mixture.

(5) Upon the completion of the addition, increase the water-bath temperature to 40 ℃ and continue to stir the reaction mixture for about one hour until the color of the mixture turns into pale yellow.

(6) Stop stirring and transfer the solution in the flask into a beaker.

(7) Cool down the mixture in an ice-water bath until the solid is completely precipitated[②].

(8) Collect the solid by vacuum filtration and wash it subsequently with cold aqueous sodium bisulfite and cold water.

(9) Press the crude 4-bromoacetanilide as dry as possible on the filter[③] and then air-dried.

(10) Weigh the dried product and calculate the percentage yield of 4-bromoacetanilide.

(11) Purify the crude 4-bromoacetanilide by recrystallization from 95% ethanol.

2.Preparation of 4-bromoaniline

(1) Mix 12.6 g of crude 4-bromoacetanilide with 35 mL of 95% ethanol in a 150 mL three-necked flask.

(2) Add several zeolites and assemble the apparatus for heating under reflux.

(3) Place 14.5 mL of concentrated hydrochloric acid in the pressure equalizing addition funnel[④].

(4) Heat the reaction mixture to reflux, and then add concentrated hydrochloric acid to the mixture slowly.

(5) After adding the concentrated hydrochloric acid, continue to reflux for 30 min.

(6) Add 50 mL of water to the resulting mixture from the top of the condenser and assemble the flask for simple distillation.

(7) Distill the solution until there is approximately 80 mL of the collected distillate.

(8) Pour the residue in the flask into a beaker containing 250 mL of ice-cold water.

(9) Add 20% sodium hydroxide solution to the resulting mixture slowly, and stir the mixture well while adding.

(10) Check the pH of the solution to ensure that it is basic. If it is not, add more aqueous base.

(11) Collect the solid precipitated by vacuum filtration, and wash it thoroughly with ice-cold water.

---

① 摇荡反应瓶，使反应物混合均匀，且固体完全溶解。
② 冰水浴冷却，确保粗产物完全析出。
③ 将布氏漏斗上的固体尽量压干。
④ 恒压滴液漏斗在使用前需要检漏。

(12) Press the crude 4-bromoanilne as dry as possible on the filter and then air-dried.

(13) Weigh the dried product and calculate the percentage yield of 4-bromoaniline.

(14) Purify the crude 4-bromoaniline by recrystallization from a mixed solvent of 1∶1 methanol/water❿.

- **Helpful Hints**

1.Preparation of 4-bromoacetanilide

(1) Hydrobromic acid is corrosive liquid, so wear latex gloves while handling it.

(2) Bromine used in this procedure is a hazardous chemical. Perform this reaction in the hood and assemble a gas-trap system to the reflux apparatus to avoid the evolution of bromine gas.

(3) The 30% hydrogen peroxide should be added slowly to avoid the evolution of the formed bromine and minimize the side reactions.

(4) TLC⓫ can be used to monitor the reaction⓬, using a mixture of 1∶1 petroleum ether/EtOAc as the developing solvent.

(5) The first filtrate produced in Step(8) and the subsequent washing solution must be quenched by sodium bisulfide⓭ before being poured into the waste container.

2.Preparation of 4-bromoaniline

(1) Concentrated hydrochloric acid is a corrosive liquid, and 30% sodium hydroxide aqueous solution is highly caustic. Wear latex gloves while handling these chemicals and wash any areas of your skin that may come in contact with them thoroughly.

(2) Concentrated hydrochloric acid should be added slowly, otherwise the reaction will be too vigorous and the solution may rush out of the reflux condenser. ⓮

(3) 4-Bromoaniline is easy to be oxidized and should be dried in the air.

- **Questions**

1. Predict the major product of the following reaction and explain your answer.

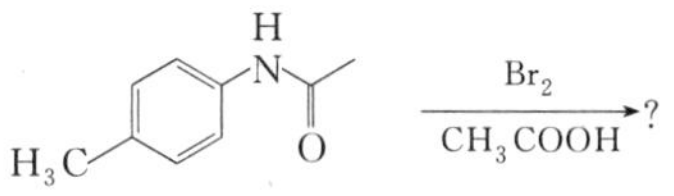

$$\text{H}_3\text{C}\text{-C}_6\text{H}_4\text{-NH-C(=O)-CH}_3 \xrightarrow[\text{CH}_3\text{COOH}]{\text{Br}_2} ?$$

2. In the work-up procedure of bromination of acetanilide, what is the purpose of each washing operation? Write down the corresponding equations of(1).

(1) Washing with cold aqueous sodium bisulfite.

(2) Washing with cold water.

3. Can saturated sodium thiosulfate solution be used to substitute saturated sodium bi-

---

❿ 粗产物可在1∶1的甲醇/水混合溶剂中重结晶提纯。
⓫ 薄层色谱技术。
⓬ 跟踪反应进程。
⓭ 亚硫酸氢钠。
⓮ 缓慢滴加浓盐酸，以免反应太过剧烈。

sulfite solution in the washing operation for the bromination of acetanilide? Why or why not?

4. In the preparation of 4-bromoaniline, why must 20% sodium hydroxide aqueous solution be added to isolate 4-bromoaniline from the aqueous acidic mixture?

5. Why is the amino group a more powerful ring activator than the amido group?

Exp.10   Pinacol Rearrangement and Photochemical Synthesis of Benzopinacol[⑮]

• **Objectives**

1. To learn the method of preparing a pinacol derivative by bimolecular photochemical reduction.

2. To learn the principle of performing a molecular rearrangement of pinacol.

3. To practice vacuum filtration and washing of solid compounds.

• **Principle**

Pinacol rearrangement[⑯] refers to the reaction, in which 2,3-dimethylbutane-2,3-diol(pinacol) is converted to 3,3-dimethyl-2-butanone(pinanone) under acid-catalyzed conditions. The corresponding reaction is shown as follows.

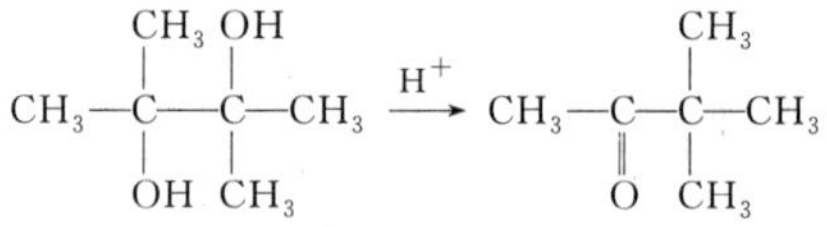

$$CH_3-\underset{\underset{OH}{|}}{\overset{\overset{CH_3}{|}}{C}}-\underset{\underset{CH_3}{|}}{\overset{\overset{OH}{|}}{C}}-CH_3 \xrightarrow{H^+} CH_3-\underset{\underset{O}{\|}}{\overset{\overset{CH_3}{|}}{C}}-\underset{\underset{CH_3}{|}}{C}-CH_3$$

The first step of pinacol rearrangement is protonation of one of the hydroxyl groups in pinacol by a protic acid, followed by loss of water to form a carbocation intermediate. Then 1,2-shift of a methyl group to yield a new carbocation, which is stabilized by an attached hydroxyl group and is more stable than the previous tertiary carbocation according to the octet rule[⑰]. Finally, loss of a proton gives pinacolone.

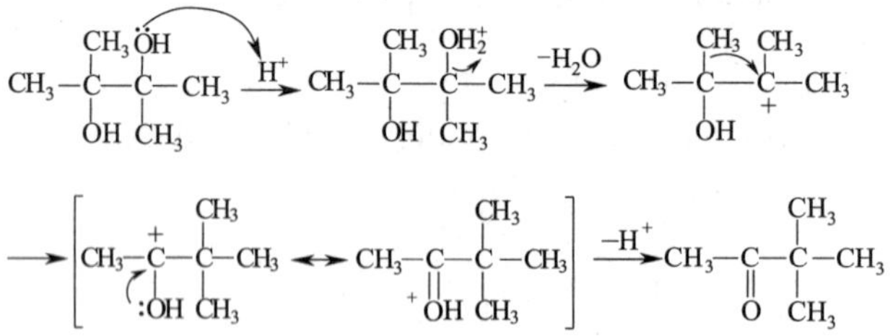

Aryl-substituted diols can also undergo the pinacol rearrangement. An example is shown below for the acid-catalyzed pinacol rearrangement of benzopinacol(1,1,2,2-tetraphenylethane-1,2-diol), which leads to the formation of benzopinacolone(2,2,2-triphenylacetophenone).

---

⑮ Reference: Allen M S, Barbara A G, Melvin L D. Microscale and Miniscale Organic Chemistry Laboratory Experiments [M]. 2nd ed. McGraw-Hill, 2004: 339-403.

⑯ 频那醇重排。

⑰ 八隅体规则。

$$\underset{\underset{OH}{|}}{\overset{\overset{Ph}{|}}{Ph-C-C-Ph}} \xrightarrow{H^+} \underset{\underset{O}{\|}}{\overset{\overset{Ph}{|}}{Ph-C-C-Ph}}$$

The photochemical reaction refers to a reaction that occurs when the reactants absorb light energy from visible or ultraviolet light[18]. Photochemical reactions can cause combination, decomposition, ionization, oxidation, reduction and other processes. The mechanism of the photochemical reactions usually involves the single electron transfer[19] and the formation of a free radical intermediate. In this experiment, benzopinacol is first synthesized by a bimolecular photochemical reduction[20] of benzophenone in 2-propanol under the light of the sun.

$$2 \; Ph\overset{O}{\underset{}{\diagdown\mkern-10mu\diagup}}Ph + \overset{OH}{\diagdown\mkern-10mu\diagup} \xrightarrow{h\nu} \underset{\underset{OH}{|}}{\overset{\overset{Ph}{|}}{Ph-C-C-Ph}} + \overset{O}{\diagdown\mkern-10mu\diagup}$$

- **Apparatus**

Apparatuses required include 50 mL Erlenmeyer flask, Büchner funnel, filter flask, 50 mL round-bottom flask, reflux condenser, and apparatus for electromagnetic stirrer.

- **Experimental Procedures**

1. Preparation of benzopinacol by photoreduction of benzophenone

(1) Place 2.0 g of benzophenone and 15 mL of 2-propanol in a 50 mL Erlenmeyer flask.

(2) Heat the mixture under a water bath to dissolve benzophenone.

(3) Add 2 drops of acetic acid[21] to the flask and cork the flask. Then, place the flask in direct sunlight for one week.

(4) Some crystals of benzopinacol is observed on the sides and bottom of the flask as the reaction goes on.

(5) Uncover the flask and collect the crystals by vacuum filtration.

(6) Wash the crystals with 3 mL of cold 2-propanol and air-dried.

(7) Weigh the final product and calculate the percentage yield of benzopinacol.

2. Preparation of benzopinacolone

(1) Place 1.5 g of benzopinacol and 15 mL of acetic acid in a 50 mL round-bottom flask.

(2) Add a few small iodine crystals[22] and a stir bar into the flask, assemble the reflux apparatus.

(3) Heat the reaction mixture at reflux for 15 min.

(4) Cool down the solution, and then add 10 mL of ethanol into the flask.

---

⑱ 光化学反应。

⑲ 单电子转移。

⑳ 双分子光化学还原。

㉑ 还原得到的苯频那醇在碱存在下很容易分解生成二苯甲酮和二苯甲醇。因而反应体系中痕量的碱在光照反应前必须除去。

㉒ 碘的加入有利于频那醇重排反应中间体碳正离子的生成。

(5) Cool down the resulting mixture in an ice-water bath until the solid precipitates completely.

(6) Collect the product by vacuum filtration.

(7) Wash the crystals with a few drops of cold ethanol and air dried.

(8) Weigh the product and calculate the percentage yield of benzopinacolone.

- **Characterization of Benzopinacolone**

1. Melting point: Determine the melting point of product and do a mixed melting point determination using a 1 : 1 mixture of benzopinacol and benzopinacolone.

2. Infrared spectroscopy: Obtain an IR spectrum of the product using either KBr and interpret the spectrum.

3. ^1H NMR spectroscopy: Obtain the ^1H NMR spectrum of the product as a solution in $CDCl_3$ and interpret the spectrum.

- **Helpful Hints**

1. 2-Propanol is flammable and volatile, make sure that there are no open flames anywhere in the laboratory. Use it in a well-ventilated area and avoid prolonged contact or inhalation.

2. Acetic acid is corrosive and suffocate, handle it in the hood and wear gloves. If any acid spills on your skin, wash it off with large amount of water.

3. In the photoreduction of benzophenone, acetic acid is added to neutralize trace amounts of base in the reaction system. Otherwise, the formed benzopinacol readily decomposes to form benzophenone and diphenylmethanol in the presence of alkali.

4. The intermediate of the photoreduction of benzophenone is free radical. Therefore, the reaction flask should be corked during the reduction process to avoid the consumption of the generated free radicals by oxygen in the air, which may slow down the reaction[①].

5. In the rearrangement of benzopinacol, iodine is added to facilitate the formation of the intermediate carbocation. Upon the completion of the reaction, iodine can be removed by washing the product with cold ethanol.

- **Questions**

1. Are there any other ways to synthesize benzopinacol from benzophenone besides the photoreduction used in this experiment.

2. Suppose the mechanism of the acid-catalyzed pinacol rearrangement of 1,1,2-triphenylethane-1,2-diol.

3. The reported melting point of benzopinacolone is 182-184 ℃ and benzopinacol is 185-186 ℃. Explain why it is necessary to do a mixed melting point determination using a 1 : 1 mixture of benzopinacol and benzopinacolone for the characterization of benzopinacolone.

4. Discuss the differences in the IR and ^1H NMR spectra of benzopinacol and benzopinacolone that are consistent with the pinacol rearrangement of benzopinacol.

---

① 在苯甲酮的双分子光化学还原反应中，应将反应瓶塞住，以免空气中的氧气进入，消耗所产生的自由基中间体，导致反应过慢。

Chapter 6

Study Guide of Some Typical Experiments

Exp.1 Preparation of Cyclohexene (Miniscale Reaction)

- **Introduction**

Alkenes are usually prepared by elimination reactions, generally involving two common reactions, i. e., dehydrohalogenation—the loss of HX from an alkyl halide and dehydration—the loss of water from an alcohol. The dehydration of alcohols is readily accomplished in the presence of an acid catalyst such as sulfuric acid or phosphoric acid. For example, cyclohexene can be prepared by cyclohexanol dehydration using conc. phosphoric acid as the catalyst. Although sulfuric acid can also be used for this reaction, phosphoric acid is a milder acid catalyst that results in higher yields of cyclohexene and fewer by-products.

$$\text{C}_6\text{H}_{11}\text{-OH} \xrightleftharpoons[\Delta]{\text{H}_3\text{PO}_4} \text{C}_6\text{H}_{10} + \text{H}_2\text{O}$$

The temperature of this reaction must be controlled carefully to avoid the competitive intermolecular dehydration of cyclohexanol to form by-products of cyclohexyl ether.

$$\text{C}_6\text{H}_{11}\text{-OH} \xrightleftharpoons[\Delta]{\text{H}_3\text{PO}_4} \text{C}_6\text{H}_{11}\text{-O-C}_6\text{H}_{11} + \text{H}_2\text{O}$$

This dehydration of cyclohexanol is an E1 mechanism and occurs by the three-step mechanism shown as follows:

$$\text{Cy-ÖH} + \text{H-O-P(=O)(OH)-OH} \rightleftharpoons \text{Cy-}\overset{+}{\text{O}}\text{H}_2 + \text{H}_2\text{PO}_4^-$$

$$\text{Cy-}\overset{+}{\text{O}}\text{H}_2 \xrightarrow{-\text{H}_2\text{O}} \text{Cy}^+ + \text{H}_2\text{O}$$

$$\text{Cy}^+ \xrightleftharpoons{-\text{H}^+} \text{C}_6\text{H}_{10}$$

The first step of the reaction is a rapid and reversible protonation of the cyclohexanol by phosphoric acid, followed by a rate-determining loss of water to generate carbocation intermediate and final loss of a proton from the neighboring carbon atom to form cyclohexene. Being a secondary alcohol, the reactivity of cyclohexanol is less than that of tertiary alcohols and can be made to react only in severe conditions(conc. H_3PO_4 and heating). Each step of the reaction pathway is reversible. In order to shift the equilibrium to the right and maximize the yield of cyclohexene, a fractional distillation is adopted to remove cyclohexene from the reaction mixture during its formation.

Miniscale organic experiments gained certain popularity in the organic laboratory during the 1980s, which employ mini-type glassware and use approximately 0.3-5 grams of starting materials. Compared with macroscale experiments, miniscale experiments are equipped with some advantages, such as environmentally friendliness, less expensiveness, shorter reaction time and higher-level safety, etc. Herein, cyclohexene is always synthesized by miniscale reaction.

- **Techniques Involved**

1. Reflux and fractional distillation.
2. Extracting and washing of liquid compounds.
3. Drying of liquid compounds.
4. Simple distillation.

- **Safety Alerts**

1. Phosphoric acid is corrosive, so gloves should be worn while handling this reagent.
2. Cyclohexene is toxic and irritating, so gloves should be worn and skin contact should be avoided while handling it.

- **Pre-lab Assignment**

1. What measures are taken to improve the yield of cyclohexene during this reaction?
2. In the Le Chatelier principle, the equilibrium of a reversible reaction can be shifted to the right by removing small molecules such as water. Why is this tactic not a good option for the dehydration of cyclohexanol?
3. What is the advantage of using phosphoric acid rather than sulfuric acid as the catalyst for the reaction?
4. Why is the head temperature of the top of the fractional column kept below 90 ℃ in the initial step of the dehydration reaction?
5. How should the ending point of this reaction be determined? Whether can the theoretical amount of water produced be used as the ending point of the reaction?
6. What is the purpose of washing the organic layer with saturated aqueous sodium chloride solution in the work-up procedure?
7. Why is it particularly important that the crude cyclohexene should be dry prior to the final distillation?
8. In the final step of distilling out the pure cyclohexene, the dried organic solution should be decaned to the distilling flask carefully to keep the drying agent anhydrous calcium chloride from falling into the flask. What is the reason for this procedure?
9. Write a flow scheme of the reaction and work-up procedure for this experiment.

Chapter 6 Study Guide of Some Typical Experiments

• **Thinking Questions**

1. What will be the expected results if HCl is substituted for H_3PO_4 as the catalyst in the dehydration of cyclohexanol? Why is the elimination favored over substitution with phosphoric acid?

2. As the catalyst of this reaction, phosphoric acid might co-distill with the crude products in the initial step of the dehydration reaction. How should phosphoric acid be removed from the reaction mixture in the work-up procedures?

3. Emulsification appears usually in the process of extracting and washing liquid compounds. What can be done to eliminate emulsification and promote the separation of the layers?

4. There are many kinds of drying agents. Why should anhydrous calcium chloride be hereby chosen as the drying agent?

5. The boiling point of cyclohexene is about 82-85 ℃, but some of the students get the final distillate less than 80℃. Please analyze the possible reason.

6. Why is an acid needed as a catalyst for the dehydration of alcohols? Give a detailed mechanism of the following reaction, using curved arrows to symbolize the flow of electrons.

7. Discuss the differences observed in the IR spectra of cyclohexanol and cyclohexene that are consistent with the dehydration in this experiment.

Exp.2 Preparation of 1-Bromobutane

• **Introduction**

The most extensively adopted method for preparing alkyl halides is to make them from alcohols, and many different methods have been developed to convert alcohols into alkyl halides. The simplest method is to treat the alcohol with HX(X= Cl,Br,or I), a representative nucleophilic substitution reaction that works best with tertiary alcohols via the S_N1 pathway. Besides, primary and secondary alcohols react much more slowly by a S_N2 mechanism, and the completion of the reaction must be guaranteed by heating. An example is the preparation of n-butyl bromide from n-butyl alcohol. In this experiment, n-butyl alcohol will be converted to n-butyl bromide via the reaction using sulfuric acid and sodium bromide. In fact, the effective reagent for the reaction is the mixture of hydrobromic acid and sulfuric acid, generated in situ by the reaction of sulfuric acid with sodium bromide. This reaction is reversible, and a large excess of the acid should be normally used to drive the equilibrium to the right. The main reactions can be shown as:

$$NaBr + H_2SO_4 \longrightarrow NaHSO_4 + HBr$$

$$HBr + n\text{-}C_4H_9OH \xrightarrow{\triangle} n\text{-}C_4H_9Br + H_2O$$

The mechanism for this reaction is known to occur via two steps: firstly, the n-butyl alcohol is protonated to give the oxonium ion, and secondly, the oxonium ion undergoes nucleophilic substitution reaction with bromide ion to form n-butyl bromide and water.

$$\text{CH}_3\text{CH}_2\text{CH}_2\text{CH}_2\ddot{\text{O}}\text{H} \xrightarrow{\text{H}-\text{Br}} \text{CH}_3\text{CH}_2\text{CH}_2\text{CH}_2\overset{+}{\text{O}}\text{H}_2 \xrightarrow{\text{Br}^-} \text{CH}_3\text{CH}_2\text{CH}_2\text{CH}_2\text{Br} + \text{H}_2\text{O}$$

The sulfuric acid plays several roles in this reaction: (1) It is a dehydrating agent that reduces the activity of water and drives the equilibrium to the right; (2) It provides an added source of hydrogen ions to protonate the hydroxyl group of alcohol and yield good leaving group water. Although the presence of sulfuric acid can promote the conversion of n-butanol to n-bromobutane, there are also several side reactions associated with sulfuric acid, which can be expressed as:

$$2\text{HBr} + \text{H}_2\text{SO}_4 \xrightarrow{\triangle} \text{Br}_2\uparrow + \text{SO}_2\uparrow + \text{H}_2\text{O}$$

$$\text{CH}_3\text{CH}_2\text{CH}_2\text{CH}_2\text{OH} \xrightarrow[\triangle]{\text{H}_2\text{SO}_4} \text{CH}_3\text{CH}_2\text{CH}=\text{CH}_2 + \text{CH}_3\text{CH}=\text{CHCH}_3$$

$$2\text{CH}_3\text{CH}_2\text{CH}_2\text{CH}_2\text{OH} \xrightarrow[\triangle]{\text{H}_2\text{SO}_4} (\text{CH}_3\text{CH}_2\text{CH}_2\text{CH}_2)_2\text{O} + \text{H}_2\text{O}$$

The occurrence of these side reactions can be minimized by controlling the reaction temperature and the concentration of sulfuric acid. Dilute sulfuric acid, a mixture of an equal volume of water and concentrated sulfuric acid, is recommended for this reaction. In this experiment, the gas-tap apparatus needs to be adopted to the reflux apparatus for absorbing the hazardous gas generated during the reaction.

- **Techniques Involved**

1. Reflux and gas absorption.
2. Extracting and washing of liquid compounds.
3. Drying of liquid compounds.
4. Simple distillation.

- **Safety Alerts**

1. Concentrated sulfuric acid is corrosive and can cause severe burns. Be extremely careful while handling it. If any concentrated sulfuric acid comes in contact with your skin, wash it off immediately with copious amounts of cold water and then with dilute sodium bicarbonate solution.

2. Alkyl halides are toxic. Some alkyl bromides are lachrymators and suspected carcinogens. Work under the hood. Wear gloves and avoid skin contact. Do not breathe alkyl halide vapors.

- **Pre-lab Assignment**

1. What measures should be taken to shift the equilibrium of the reaction to the right?
2. Why is HBr generated in situ from NaBr and H_2SO_4 rather than just by using concentrated HBr in the conversion of n-butyl alcohol to n-butyl bromide?
3. What methods are usually used to help determine which layer is the aqueous layer and which is the organic layer when extracting or washing organic liquid compounds?
4. Dibutyl ether, 1-butene and 2-butene are usually obtained as by-products in this exper-

iment. What measures should be taken to remove them in the work-up procedures? Write down the corresponding equations.

5. What is the purpose of each washing operation in the workup procedures of this experiment?

(1) Washing the organic layer with concentrated H_2SO_4.

(2) Washing the organic layer with water.

(3) Washing the organic layer with 10% Na_2CO_3.

(4) Washing the organic layer with water again.

6. The crude n-butyl bromide is dried with anhydrous calcium chloride. Can solid sodium hydroxide or potassium hydroxide be used for this purpose? Explain the reasons.

7. Why must the crude n-butyl bromide be dried completely prior to the final distillation?

8. Specify the absorption associated with the C—Br bond in n-butyl bromide and the O—H bond in n-butyl alcohol in the IR spectrum.

9. Draft a flow scheme of the reaction and work-up procedure for this experiment.

- **Thinking Questions**

1. Why must the gas-trap apparatus be used in this reaction? How should the right absorbent be chosen for gas-trap apparatus?

2. In the preparation of n-butyl bromide, what will happen if the concentration of the sulfuric acid is too high or too low?

3. n-Butanol can also be converted into n-butyl bromide by the treatment with phosphorus tribromide. List health and safety considerations for phosphorus tribromide.

4. During the work-up procedures, the organic layer turns to be reddish-brown after being washed with concentrated sulfuric acid. What is the reason? What should be done for its treatment?

5. When the distillate is transferred to a separatory funnel after the crude n-butyl bromide is distilled out with water and other impurities, the organic layer is normally in the bottom layer, but sometimes it is in the top layer. Explain the reason.

6. Crude n-butyl bromide can be purified preliminarily by the first simple distillation to separate from most of the unreacted 1-butanol and other by-products. When the following apparatus is used for the distillation, is there any other method to determine the endpoint of the distillation besides checking the miscibility of the distillate in water?

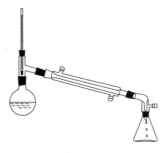

7. Discuss the differences observed in the IR spectrum of *n*-butyl alcohol and *n*-butyl bromide that are consistent with the conversion in this experiment.

8. Predict all of the possible products when 3-methyl-2-butanol is treated with HBr and propose the corresponding mechanism of this reaction.

9. Treating 1-butanol with sulfuric acid leads to the formation of 1-butylbisulfate and water, and the main reaction can be shown as:

$$CH_3(CH_2)_2CH_2OH + H_2SO_4 \rightleftharpoons CH_3(CH_2)_2CH_2OSO_3H + H_2O$$
$$\text{1-Butyl bisulfate}$$

1-Butylbisulfate suffers elimination on heating to give a mixture of alkenes. Provide the structures of the alkenes produced from 1-butyl bisulfate.

Exp.3 Preparation of Acetyl Ferrocene and Column Chromatography

• **Introduction**

Friedel-Crafts acylation reaction is an example of electrophilic substitution reaction, involving the substitution of a hydrogen atom on the aromatic ring by an acyl cation, which introduces an acyl group to the ring. Carboxylic acid chloride and carboxylic acid anhydride are the two most used acylating reagents for Friedel-Crafts acylation reaction.

Ferrocene is a well-known sandwich coordinate complex, with two cyclopentadienyl rings coordinating to a ferrous ion. The cyclopentadienyl rings filled with 6 π electrons are highly electron rich and aromatic, and can therefore undergo electrophilic substitution reaction with high reactivity. For example, ferrocene can undergo Friedel-Crafts acylation reaction with acetic anhydride in a milder condition to yield acetyl ferrocene using phosphoric acid as the catalyst. In order to minimize the formation of the by-product of 1,1-diacetylferrocene and prevent the ferrocene from being oxidized, phosphoric acid should be added dropwise to the cold solution of acetic anhydride and ferrocene while keeping the reaction temperature at desired range.

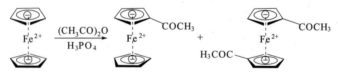

The mechanism of Friedel-Crafts acylation reaction involves the addition of the electrophilic acyl cation to one of the carbon atoms on the cyclopentadienyl ring to yield a nonaromatic carbocation intermediate. A base removes H^+ from the intermediate and the acetyl ferrocene is produced, as two electrons from the C—H bond move to re-form the aromatic cyclopentadienyl ring. In the acetylation of ferrocene, the reactive electrophile is an acetyl cation generated by the reaction of acetic anhydride with phosphoric acid.

Being a solid-liquid adsorption chromatography, column chromatography can efficiently separate and purify compounds in a larger scale from milligrams to hundreds of grams in organic laboratory. Column chromatography is based on the different adsorption capacities of each component in a mixture with the adsorbent. In column chromatography, the stationary phase is packed into a column and the sample is adsorbed on the adsorbent at the top of the column. When the eluting solvent is added from the top of the column, the components of the sample will begin to move down the column with the eluting solvent. The interaction of the individual component with the stationary phase and the mobile phase determines the rate at which the component elutes from the column. In general, the more weakly the component is absorbed, the more rapidly the component will be eluted. Thus, with a polar adsorbent such as silica gel or aluminum oxide, the less polar component travels faster down the column than the more polar components, eventually leading to the separation of the components from the column in order. The proper choice of elution solvents is the key to a successful column chromatography separation. Thin-layer chromatography can provide a guide to determine the most suitable eluting solvent system for column chromatography separation. Generally, the less polar solvent is first used to elute the component with the less polar, and a more polar solvent or combination of mixed solvents is then applied to elute the polar component.

In this experiment, column chromatography is applied to separate the desired acetyl ferrone from unreacted starting materials ferrocene and the by-product 1,1'-diacetylferrocene based on differences in the adsorption capacity of these three substances on silica gel. All the substances to be separated in this experiment are highly colored, making it easy for the column chromatography separation to observe the separation visually.

- **Techniques Involved**

1. Reflux and mechanical stirring.
2. Addition liquid compounds through a addition funnel with pressure equilibrating arm.
3. Cooling with an ice-water bath.
4. Vacuum filtration and washing of solid compounds.
5. Column chromatography.
6. Rotary evaporation.

- **Safety Alerts**

1. Phosphoric acid is corrosive. Wear gloves while handling this reagent.
2. Acetic anhydride is corrosive and irritating. Wear gloves and avoid skin contact while handling it, so as to minimize exposure to it. Perform this reaction in the hood. Do not allow acetic anhydride to come in contact with your skin. If so, flush the affected area with copious amounts of water.
3. Petroleum ether is a highly volatile and flammable mixture of low-boiling point hydrocarbons. During the packing and development of the chromatographic column, make sure that there are no flames in the vicinity.
4. As a lot of volatile organic solvent and small particles of silica gel are used in column chromatography, special safety precautions should be taken.

- **Pre-lab Assignment**

1. Explain the reason for the higher reactivity of ferrocene than that of benzene toward

electrophilic substitution reaction.

2. Write the acetylation mechanism of ferrocene by acetic anhydride using phosphoric acid as the catalyst.

3. The directive effects of substituents are important in the electrophilic aromatic substitution reaction. Try to explain why the acylation never occurs more than once on an aromatic ring.

4. Ferrocene is more susceptible to electrophilic substitution reaction than benzene. Why is it difficult to get a desired product when ferrocene is nitrated with a mixture of concentrated nitric and sulfuric acids.

5. In this experiment, why must phosphoric acid be added dropwise to the cold solution of acetic anhydride and ferrocene?

6. Compare the relative polarities of ferrocene, acetyl ferrocene and 1,1′-diacetylferrocene.

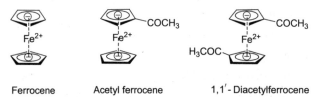

Ferrocene   Acetyl ferrocene   1,1′- Diacetylferrocene

7. A mixture of the following three compounds (a)-(c) is separated by column chromatography, using neutral aluminum oxide as the adsorbent and petroleum ether as the eluting solvent. Predict the order following which compounds (a)-(c) will be eluted from the column.

(a) 4-methylphenyl-CH(CH$_3$)OH   (b) 4-methylbenzoic acid   (c) 4-methylcumene

8. Design a flow scheme for the work-up procedures used in this experiment.

• **Thinking Questions**

1. What feature of the acylation of ferrocene that makes it particularly attractive for column chromatography separation?

2. In this experiment, can acetyl ferrocene be purified by recrystallization? Compare the advantages and disadvantages of recrystallization and column chromatography separation when they are used to purify the acetylferrocene.

3. In the elution process of the sample, different absorbed components with progressively more polar eluting solvent can be eluted. Why must changing eluting solvents follow the order from a less polar to a more polar solvent? Explain the reason.

4. Given that ferrocene can be eluted by petroleum ether, why is a mixture of 1∶1 petroleum ether/methylene chloride used to elute the acetylferrocene, and why does diacetyl ferrocene have to be eluted by methylene chloride in the column chromatography separation of the mixture consisting of these three substances?

5. Library project: search for other methods and figure out which is more eco-friendly and highly selective for the acetylation of ferrocene?

6. Discuss the differences observed in the IR and ¹H NMR spectrum of acetyl ferrocene and ferrocene that are consistent with the conversion in this experiment.

Exp.4　Preparation of Acetylaniline

● **Introduction**

The amino function on an aromatic ring is an electron-donating group that activates the aromatic ring toward electrophilic aromatic substitution reactions. However, the high reactivity of the amino group on the aromatic rings may sometimes be a drawback. For example, the halogenation reaction of aniline takes place rapidly and polysubstituted products are usually formed. It is hard to stop this reaction at the monosubstitution stage to form monosubstituted products. Besides, the basic and nucleophilic amino group can directly react with acids and other electrophilic reagents, thus affecting the occurrence of the desired reaction. Therefore, sometimes it is necessary to protect the aryl amino group to ensure the desired reactions.

In general, placing a protecting group on a reactive functional group alters the reactivity of that group and makes it unreactive toward certain reagents. The protecting group must be stable in normal reaction conditions and can be removed easily when the desired reaction has been finished. The aryl amino group(—NH_2) is usually protected as an aryl amide(—NHCOR) by acylation reaction. The amide group is moderately activated with *ortho-* and *para*-directing, which can also be used to block the functionality of active amino group and thus improve the regioselectivity and chemoselectivity of the desired reactions. For example, the amide group on the aromatic ring makes the monohalogenation of acetylaniline possible and directs the halogenation preferentially to the para position due to the steric hindrance. The acetylation of the aryl amino group is reversible and the acyl group will be removed to regenerate the free amino group via acid hydrolysis upon the completion of the desired reactions.

The acetylation is usually accomplished by typical acyl nucleophilic substitution reaction between an amino group with a carboxylic acid derivative, such as acyl halides or acid anhydrides. Acyl chloride reacts vigorously with amino group and thus makes the reaction difficult to control. Acid anhydride is a common acylation reagent and preferred for a laboratory synthesis, which, however, is expensive and sensitive to water, with its application in undergraduate laboratory thereby limited. Moreover, the diacetylation reaction may sometimes occur accompanied with the related acylation when acid anhydrides are used to protect the amino groups.

$$\underset{}{\text{C}_6\text{H}_5\text{NH}_2} \xrightarrow[\text{RCOONa}]{(RCO)_2O} \underset{}{\text{C}_6\text{H}_5\text{NHCOR}} \xrightarrow{(RCO)_2O} \underset{\text{By-product}}{\text{C}_6\text{H}_5\text{N(COR)}_2}$$

Carboxylic acids can also be used for the acylation of the amino group under certain condition. Although the carboxylic acids are less reactive than carboxylic acid derivatives and the acylation of the amino group with carboxylic acids always requires more time, the procedure is still endowed with certain commercial interest. In this experiment, acetylaniline is prepared from aniline using acetic acid as the acyl reagent in good yield. The reaction is shown as below:

$$\text{C}_6\text{H}_5\text{NH}_2 + \text{CH}_3\text{COOH} \xrightleftharpoons{\text{Zn}} \text{C}_6\text{H}_5\text{NHCOCH}_3 + \text{H}_2\text{O}$$

As this reaction is reversible, the product water is removed once it is formed by fractional distillation to allow this reaction to go forward.

- **Techniques Involved**

1. Reflux and fractional distillation.

2. Vacuum filtration and washing.

3. Recrystallization.

- **Safety Alerts**

1. Aniline is a highly toxic irritant. Acetic anhydride is a corrosive lachrymator. Avoid skin contact and inhalation of the vapors.

2. Acetic acid is corrosive and suffocate, so handle it in the hood and wear gloves. If any acid spills on your skin, wash it off with large amount of water.

- **Pre-lab Assignment**

1. Why are amide groups(—NHCOR) less strongly activating and less basic than amino groups(—NH$_2$)?

2. The acetlyation of aniline by acetic acid is reversible, so what measures should be taken to maximize the yield of acetylaniline in this experiment?

3. Why is fractional distillation used in this reaction?

4. What is the purpose of adding zinc powder to the reaction mixture?

5. Why must the temperature of distilling vapor be controlled at 105 ℃ in the initial step of the acetylation reaction?

6. In addition to the sudden drop of temperature, are there any other observed phenomena that indicate the completion of the acylation reaction?

7. Recrystallization is a rather important method to purify the solid compounds. What are the requirements of a suitable solvent for recrystallization? What steps are involved in recrystallization operation?

8. During the recrystallization operation, which step are the insoluble and soluble impurities removed at respectively?

9. Specify the absorption associated with the N—H bond, C=O double bond, and the aromatic ring in the functional group region of the IR spectrum for acetylaniline.

10. Draft a flow scheme for each reaction and work-up procedure in the synthesis.

- **Thinking Questions**

1. After the acetylation is finished, what is the purpose of pouring the reaction mixture

in the flask into water while it is still hot? What will happen if the mixture is poured after being cooled down?

2. How many milliliters of water should be obtained at the end of the reaction according to the theoretical calculation? More liquid is usually collected in the receiving flask than the theoretical amount. Explain the reason.

3. In recrystallization operation, if the hot saturated solution is colorless, is it still necessary to add activated carbon to the hot solution? Explain the reason.

4. In the recrystallization of a white solid compound, some white solid substances are found in the Büchner funnel besides the black activated carbon after the hot filtration. Explain the reason.

5. Discuss the differences observed in the IR spectrum of acetylaniline and aniline that are consistent with acylation occurring in this experiment.

6. Here is the ^{1}H NMR of acetylaniline(CDCl$_3$ as the solvent). Assign various resonances to the hydrogen nuclei responsible for them.

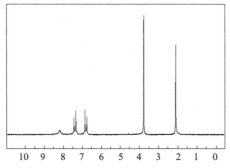

7. Give a stepwise reaction mechanism for the reaction of aniline with acetic anhydride, and use curved arrows to symbolize the flow of electrons.

8. Many applications of the aniline acylation in organic synthesis have been found. Try to summarize them and complete the following transformations.

Exp.5 Preparation of 2-Methyl-2-butanol

• **Introduction**

Organometallic compounds are substances that contain carbon-metal(C—M) bonds, in which the metal, M=Li, Na, Mg, Pd, or other transition elements. The carbon-metal covalent bonds in organometallic compounds are polar, and the carbon atoms bearing partial negative charge are electron-rich, or Lewis basic. Such carbon atoms then serve as nucleophile in many organic reactions. Organomagnesium compounds(M = Mg), commonly known as Grignard reagents, are important examples of organometallics. Grignard reagents, R—MgX, may be formed by reacting magnesium with alkyl or aryl halides.

$$R(Ar)-X + Mg \xrightarrow{\text{Dry ether}} R(Ar)-MgX \quad X=Cl, Br, I$$

The preparation of Grignard reagents theoretically requires equivalent amounts of alkyl halides and magnesium, but a slight excess of magnesium is usually used. Sometimes, the reactions of magnesium with alkyl or aryl halides are difficult to initiate, especially with unreactive halides. Thus, a small crystal of iodine can be added to the reaction mixture to initiate the reactions. Alternatively, the mixture of magnesium and alkyl halides in ether can be placed in a warm water-bath to accelerate the initiation of the reaction.

The solvent plays a key role in the Grignard reactions. Given that the basic oxygen atoms of the ether can form soluble complexes with the magnesium and thus help to stabilize the Grignard reagents, anhydrous ether solvents such as diethyl ether or tetrahydrofuran (THF) are critical for the efficient preparation of the Grignard reagents.

Some side reactions may occur during the formation of Grignard reagents just as follows:

$$RMgX + H_2O \longrightarrow RH + Mg(OH)X$$
$$RMgX + 1/2 O_2 \longrightarrow ROMgX \xrightarrow{H_2O} ROH + Mg(OH)X$$
$$RMgX + CO_2 \longrightarrow RCO_2MgX \xrightarrow{H_2O} RCOOH + Mg(OH)X$$
$$RMgX + RX \longrightarrow R-R + MgX_2$$

Grignard reagents can react with water, which requires that all reagents, solvents, and apparatus used for Grignard reactions must be dry. The reactions of the Grignard reagent with oxygen and carbon dioxide can be avoided by performing the reaction under an inert atmosphere such as nitrogen (N_2) or argon (Ar). The last coupling reaction of alkyl halides with the corresponding Grignard reagents can be minimized by decreasing the localized concentrations of the alkyl halides, which is accomplished by stirring and slowly adding the dilute ether solution of halides to the suspension of magnesium in dry ether.

Considering their nucleophilic character, Grignard reagents can react with many kinds of electrophiles, such as aldehydes and ketones, epoxides and esters to produce new carbon-carbon bonds. The nucleophilic addition reactions of Grignard reagents with aldehydes or ketones are commonly used to extend the carbon chain and synthesize various kinds of secondary or tertiary alcohols. This reaction includes two steps: the first is the addition reaction to give alkoxide intermediate and the second is to convert it to alcohol upon protonation with acid.

The experiment to be performed involves the initial preparation of ethylmagnesium bromide from ethyl bromide with magnesium metal in an anhydrous dry ether solvent, then ethylmagnesium bromide react with acetone, followed by an acid workup to produce the final product of 2-methyl-2-butanol. Here, anhydrous diethyl ether with low boiling point and very high vapor pressure is used as the solvent for the Grignard reactions to effectively exclude most of the air from the reaction vessels, which leads to a relatively oxygen-free reaction condition.

Step 1: $CH_3CH_2\text{—}Br + Mg \xrightarrow{\text{Dry ether}} CH_3CH_2\text{—}MgBr$

Step 2: $CH_3CH_2\text{—}MgBr + O\!\!=\!\!C(CH_3)_2 \xrightarrow{\text{Dry ether}} H_3C\underset{\underset{OMgBr}{|}}{\overset{\overset{CH_3}{|}}{C}}CH_2CH_3$

Step 3: $H_3C\underset{\underset{OMgBr}{|}}{\overset{\overset{CH_3}{|}}{C}}CH_2CH_3 \xrightarrow{H_3^+O} H_3C\underset{\underset{OH}{|}}{\overset{\overset{CH_3}{|}}{C}}CH_2CH_3 + Mg(OH)Br$

- **Techniques Involved**

1. Reflux and magnetic stirring.

2. Adding liquid compounds through an addition funnel with pressure equilibrating arm.

3. Cooling under a water bath.

4. Extracting, washing and drying of liquid compounds.

5. Simple distillation.

- **Safety Alerts**

1. Diethyl ether and ethyl bromide are extremely flammable and volatile, so be sure that there are no open flames anywhere in the laboratory. Use the low boiling point liquid in a well-ventilated area and avoid prolonged contact or inhalation.

2. Concentrated sulfuric acid is corrosive and can cause severe burns. Be very careful while handling it. If any concentrated sulfuric acid comes in contact with your skin, immediately wash it off with copious amounts of cold water first and then with dilute sodium bicarbonate solution.

3. Alkyl halides are toxic, and some alkyl bromides are lachrymators and suspected carcinogens, so work in the hood. Wear gloves and avoid skin contact. Do not breathe in alkyl halide vapors.

- **Pre-lab Assignment**

1. Grignard reagents are air-sensitive and water-sensitive. What measures should be taken to create these conditions in the preparation of ethylmagnesium bromide?

2. Ethanol is often present in solvent-grade diethyl ether. If the solvent-grade diethyl ether rather than anhydrous is used for the preparation of ethylmagnesium bromide, what impact will the ethanol have on the formation of the Grignard reagents?

3. Why can the remaining ethereal solution of ethyl bromide not be added to the reaction mixture until the Grignard reaction has been initiated?

4. Why is it important to add the ethereal solution of the ethyl bromide dropwise to the suspension magnesium turnings in the diethyl ether?

5. Why can anhydrous calcium chloride not be used to dry crude 2-methyl-2-butanol prior to the final distillation?

6. Why must the crude 2-methyl-2-butanol be dried thoroughly before the final distillation?

7. Specify the absorption associated with the C=O bond in acetone and the O—H bond in 2-methyl-2-butanol in the IR spectrum.

8. Draft a flow scheme of the reaction and work-up procedure for this experiment.

- **Thinking Questions**

1. Why can a small crystal of iodine initiate the Grignard reaction?

2. Why is it wrong to begin the addition of the ethereal solution of acetone to the solution of ethylmagnesium bromide before the latter has been placed in an ice-water bath?

3. During the preparation of ethylmagnesium bromide, small quantities of butane are formed. Explain how the undesired butane is formed and how its formation can be minimized.

4. After the addition of dilute sulfuric acid, a clear solution should be obtained for this experiment. However, some students found that there were still some white precipitation in the reaction flask, which might affect the next work-up procedure. What is the precipitation? What should be done next?

5. What is the purpose of each washing and extracting operation in this experiment?
(1) Wash the organic layer with 8 mL 10% sodium carbonate.
(2) Extract the water layer with ether(6 mL×2).

6. The boiling point of 2-methyl-2-butanol is about 100-102 ℃. However, certain distillate is obtained below 95℃. Explain the possible reason.

7. The reaction of 5-bromo-1-pentanol with magnesium in anhydrous ether failed to give the expected Grignard reagent, 5-hydroxy-1-pentylmagnesium bromide. Provide an explanation for the failure of this reaction. What organic product is formed instead?

8. 2-Methyl-4-pentene-2-ol can be prepared by the reaction of allyl Grignard reagent with the corresponding ketone. Please write down the related reaction equations. In the laboratory, the saturated solution of NH_4Cl is often used to replace HCl to quench the reaction finally. Explain the possible reason.

9. How might primary, secondary, and tertiary alcohols be prepared from a Grignard reagent and a suitable carbonyl-containing compounds besides aldehydes and ketones? Write down the chemical reactions for these preparations using any starting materials as expected.

10. Discuss the differences observed in the IR and ^1H NMR spectra of acetone and 2-methyl-2-butanol that are consistent with the conversion of a ketone into a tertiary alcohol in this experiment.

Exp.6 Preparation of Benzil and Thin-layer Chromatography

- **Introduction**

1,2-Diphenylethane-1,2-dione, also known as benzil, is one of the most common α-diketones, generally utilized as an important intermediate in pharmaceuticals and organic synthesis. It is also a potent inhibitor of human carboxylesterases involved in the hydrolysis of many drugs used clinically. In addition, benzil is also a commonly used photo-initiator for polymerization and has many applications in materials science.

Benzil can be easily prepared by oxidation of benzoin with a variety of oxidizing agents. Nitric acid is the first oxidizer used in the oxidation of benzoin to benzil. Although this method is easy to operate, the reaction is somewhat violent and not easy to control. Moreover, during the reaction, nitric acid is reduced to form nitrogen oxides(NO_x), which are all environmental contaminants. With the continuous development of organic synthesis technology, a large number of inorganic and organic oxidizers have been used for the oxidation of benzoin, including copper sulfate in pyridine, $Cu(OAc)_2/NH_4NO_3$, $CrO_3 \cdot CH_3NH_2 \cdot HCl$, $(PhCO)_2/MeCN$, $(COCl)_2/CH_2Cl_2$ and some metallorganic oxidants, etc. Although these oxidants are effective for the conversion of benzoin to benzil, the reaction conditions of some methods are harsh or cause hazards to human health as well as the environment, and some methods may be subject to tedious work-up procedure, expensive regents or low conversion ratio. The greener replacement of oxidants may be one of the most active study areas for the oxidation of benzoin. Some green oxidizing agents have been used for the preparation of benzil. For example, benzil can be synthesized by the oxidation of benzoin in air using organometallic complex [Co(Salen)] as the catalyst. This reaction is performed in milder conditions with a high yield, and the catalyst can be recycled and reused several times.

Considering the operability and consistent results, in this experiment, ferric trichloride is used as the oxidizing agent to oxidize benzoin to benzyl. Here, acetic acid is not only used as a solvent, but also used to provide an acidic medium for this conversion to inhibit the hydrolysis of ferric chloride. The main reaction is shown as below:

$$\underset{H}{\underset{|}{\underset{\underset{\|}{O}}{Ph-C}-\underset{\underset{OH}{|}}{C}-Ph}} + FeCl_3 \xrightarrow[\Delta]{CH_3COOH} Ph-\underset{\underset{\|}{O}}{C}-\underset{\underset{\|}{O}}{C}-Ph + FeCl_2 + HCl$$

It is meaningful to accurately determine the end point of the reaction in organic laboratory. Too long reaction time will result in the waste of time and energy. Furthermore, some changes or decomposition of the product may also occur with a longer reaction time, thus affecting the yield and purity of the final product.

Thin-layer chromatography is a simple, inexpensive and efficient method for rapid analysis of small quantities of samples, especially used in organic laboratory to identify the components in a mixture, to separate compounds at a scale between microgram to gram, to monitor the progress of a reaction and to serve as a forerunner of column chromatography. In this experiment, thin-layer chromatography analysis is used to help accurately determine the end point of the oxidation reaction by observing the gradual disappearance of the benzoin spot in the reaction mixture.

- **Techniques Involved**

1. Reflux.
2. Vacuum filtration and washing of solid compounds.

3. Thin-layer chromatography and fluorescence visualization technique.

4. Recrystallization.

• **Safety Alerts**

1. Acetic acid is corrosive and suffocate, so handle it in the hood and wear gloves. If any acid spills on your skin, wash it off with large amount of water.

2. Ferric trichloride is corrosive, so wear gloves while handling it. In addition, Ferric chloride can deliquesce easily, so handle it as fast as possible.

• **Pre-lab Assignment**

1. What is the purpose of adding acetic acid to the reaction mixture in this experiment?

2. When carrying out a vacuum filtration, what problem will occur if you turn off the aspirator before closing the stopcock on the trap?

3. List the main steps of recrystallization and briefly explain the purpose of each step.

4. In thin-layer chromatography separation, how should the R_f values of different components be calculated? List the factors that will affect R_f value of a component.

5. Compare the polarity of benzoin and benzil. If dichloromethane is used as the eluting solvent to separate these two compounds, which one is going to have a higher R_f value than the other?

6. Many organic compounds are colorless and the TLC analysis cannot be observed directly. How many indirect visualization techniques can be used in TLC analysis for colorless compounds? Briefly describe the principle of these visualization techniques.

7. In a TLC separation, why must the sample applied on the TLC plate be higher than the surface of the eluting solvent in the developing chamber?

8. Draft a flow scheme for each reaction and work-up procedure of this experiment.

• **Thinking Questions**

1. In addition to ferric chloride, what other oxidants can be used to oxidize benzoin to form benzil? What are the advantages and disadvantages of these oxidants?

2. Summarize the application of TLC separation technique in organic synthesis.

3. If the following two compounds are separated by TLC separation using the mixture solution of benzene and ethyl acetate in the ratio of 9∶1 as the eluting solvent, which of the two compounds has a smaller R_f value?

4. In this experiment, dichloromethane is used as the eluting solvent for the TLC analysis to monitor the reaction. If the eluting solvent is replaced by methanol, what will be the impact on the R_f value of benzoin and benzil?

5. Discuss the differences observed in the IR spectrum of benzoin and benzil that are con-

sistent with oxidation occurring in this experiment.

6. Benzoin is a white solid while benzil is a green yellow. Explain the reason for this color change.

7. TLC can also be used to separate the *cis-* or *trans-* isomers of a compound. If the two isomers of azobenzene are separated by thin-layer chromatography, which isomer has a smaller R_f value?

$$\underset{(1)}{\underset{C_6H_5}{\diagdown}N{=}\ddot{N}\underset{C_6H_5}{\diagdown}} \qquad \underset{(2)}{\underset{C_6H_5}{\diagdown}\ddot{N}{=}\ddot{N}\underset{}{\diagup}C_6H_5}$$

Exp.7 Preparation of Cinnamic Acid

• **Introduction**

Aldehydes or ketones containing an α-H can undergo a base-catalyzed carbonyl condensation reaction, i.e., aldol condensation reaction. In this reaction, one partner is converted into a nucleophilic enolate ion and added to the electrophilic carbonyl group of the second partner. The product of the aldol condensation reaction is always β-hydroxy aldehydes or ketones, which can then be easily dehydrated to yield α,β-unsaturated products. The great value of carbonyl condensations is that they are one of the general methods for forming C—C bonds, thereby making it possible to build larger molecules from smaller precursors. Some aldol-type condensations can also occur between aldehydes or ketones with other different carbonyl components, such as esters and acid anhydrides.

The Perkin reaction is a typical aldol-type condensation in which an aromatic aldehyde (ArCHO) reacts with an acid anhydride, $(RCH_2CO)_2O$ in the presence of alkali salts as the base catalyst, of which carboxylate salts and carbonate are often used. Such condensation reaction generates *β-aryl-α,β*-unsaturated carboxylic acid(ArCH=CRCOOH). Herein, aromatic aldehydes containing no α-H are good electrophilic acceptors, while acid anhydrides containing α-H can be easily converted into corresponding nucleophilic enolate ions when treated with a base. The formed enolate ions act as the nucleophile and are then added to the electrophilic carbonyl group of aromatic aldehyde, then a aldol-type condensation reaction is likely to be successful.

$$ArCHO + (RCH_2CO)_2O \xrightarrow[\triangle]{\text{Alkali salts}} ArCH{=}CRCOOH + RCH_2COOH$$

Cinnamic acid can be synthesized from benzaldehyde and acetic anhydride by Perkin reaction at high temperature, with potassium acetate taken as a basic catalyst. The reaction is indicated as below:

$$\text{C}_6\text{H}_5\text{—CHO} + (CH_3CO)_2O \xrightarrow[\triangle]{\text{KAc}} \text{C}_6\text{H}_5\text{—CH=CHCOOH} + CH_3COOH$$

The mechanism of this reaction is as follows:

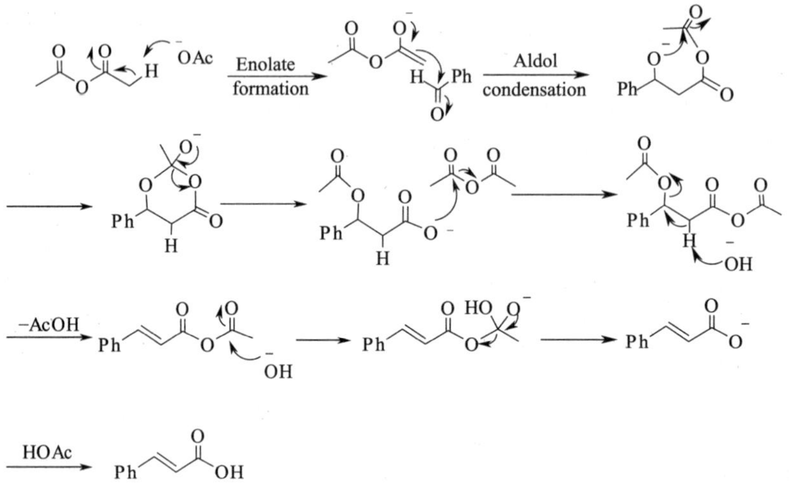

$$\text{CH}_3\text{CO-O-COCH}_3 \xrightarrow[\text{formation}]{\text{Enolate}} \xrightarrow[\text{condensation}]{\text{Aldol}} \cdots \xrightarrow{-\text{AcOH}} \cdots \xrightarrow{\text{HOAc}} \text{Ph-CH=CH-COOH}$$

The Perkin reaction between benzaldehyde and acetic anhydride generally needs to be carried out at high temperature. However, at the same time, some side reactions, such as decarboxylation and polymerization, may occur if the reaction mixture is heated at high temperature for a long time. These side reactions lead to the formation of some resin-like impurities, which are the trouble for the following work-up procedure. In this case, the temperature of this reaction should be controlled between 160-170 ℃ to minimize the occurrence of side reactions.

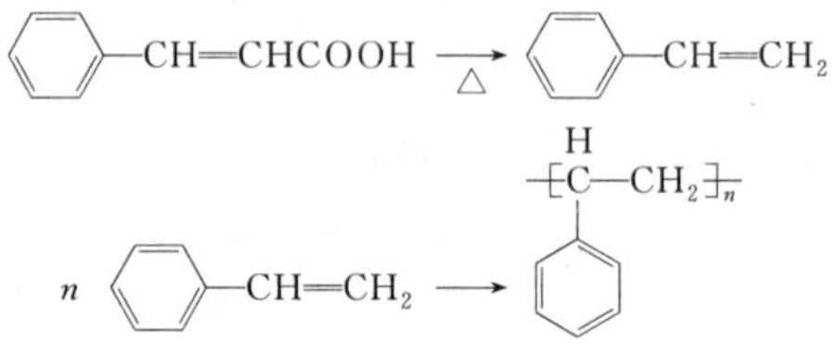

$$\text{Ph-CH=CHCOOH} \xrightarrow{\Delta} \text{Ph-CH=CH}_2$$

$$n \text{ Ph-CH=CH}_2 \longrightarrow \text{-[CH(Ph)-CH}_2\text{]}_n\text{-}$$

• **Techniques Involved**

1. Reflux and steam distillation.
2. Vacuum filtration and washing of solid compounds.
3. Recrystallization.
4. Melting-point measurement.

• **Safety Alerts**

1. Acetic anhydride is corrosive and irritating. Wear gloves and avoid skin contact while handling it, perform this reaction in the hood. Do not allow acetic anhydride to come in contact with your skin. If it does, flush the affected area with copious amounts of water.

2. Benzadehyde is toxic and irritating, so wear gloves and avoid skin contact while handling it. Benzadehyde is easily oxidized to form benzoic acid. It should be redistilled before use.

3. *trans*-Cinnamic acid is an irritant, so wear gloves and avoid skin contact.

• **Pre-lab Assignment**

1. What kinds of aldehydes and anhydrides can be used to perform the Perkin reaction?

2. What measurements should be taken to minimize the side reactions in this experiment?

3. What might be the impact of too long reaction time and too high reaction temperature in the initial step of this reaction?

4. Why is the air condenser applied in the initial step of this reaction? What may happen if the air condenser is replaced by the reflux condenser?

5. What is the principle of steam distillation? Why is steam distillation applied in this experiment?

6. Why is it necessary to adjust the pH value of the reaction mixture to 8 prior to steam distillation?

7. What substances are removed by activated carbon after the mixture with activated carbon is heated for several minutes?

8. To determine the melting point, why must the heating rate be reduced gradually when the temperature is close to the melting point of the compounds being measured?

9. Draft a flow scheme of the reaction and work-up procedure for this experiment.

- **Thinking Questions**

1. What is the influence on the reaction if there is a trace of benzoic acid in the starting material of benzaldehyde? What should be done with the benzaldehyde if it has been stored for a long time?

2. What undesired reaction might occur if the pH is adjusted with strongly basic aqueous solution of sodium hydroxide rather than sodium carbonate prior to the steam distillation?

3. What can be obtained from the reaction of benzaldehyde with propanoic anhydride in the presence of anhydrous potassium carbonate? Write down the corresponding reaction equation.

4. When will steam distillation be used to separate and purify reaction mixtures?

5. When an acidic aqueous solution of the following three compounds is steam-distilled, which compound will be distilled out together with water? Why?

$H_3C-\langle\rangle-NO_2$ $H_3C-\langle\rangle-NH_2$ (pyridine)

(a)   (b)   (c)

6. Try to analyze the factors that might affect the yield and purity of the cinnamic acid in this experiment.

7. Specify the absorptions associated with the benzene ring and —COOH function group in the functional group region of the IR spectrum.

8. Give the structures of *trans-* and *cis-* isomers of cinnamic acid. What methods can be used to distinguish these two kinds of isomers?

Exp.8 Preparation of Benzoic Acid and Ethyl Benzoate

- **Introduction**

Microwave-assisted reaction is a new energy input method, which makes the organic synthesis more environmentally-friendly and "greener". In recent years, microwave apparatus has been widely applied in organic synthesis. Microwave-assisted reaction is instructive compared with the conventional direct heating method in term of its significant advantages, including greatly shortened reaction time, decreased exposure to toxic chemicals, decreased thermal degradation, lower cost, and easiness of working up the reaction mixture. Sometimes, microwave irradiation can also increase the yield of the product.

Simple alcohols are often used as reactants to synthesize carboxylic acid in the undergraduate organic laboratory because many alcohols are relatively cheap, nonhazardous and easy to perform. In this experiment, benzoic acid is prepared by the microwave-assisted oxidation of benzyl alcohol using potassium permanganate as the oxidizing agent. After benzoic acid is obtained, it will further perform Fischer esterification with ethanol in the presence of concentrated sulfuric acid to give ethyl benzoate. The reactions are shown as below:

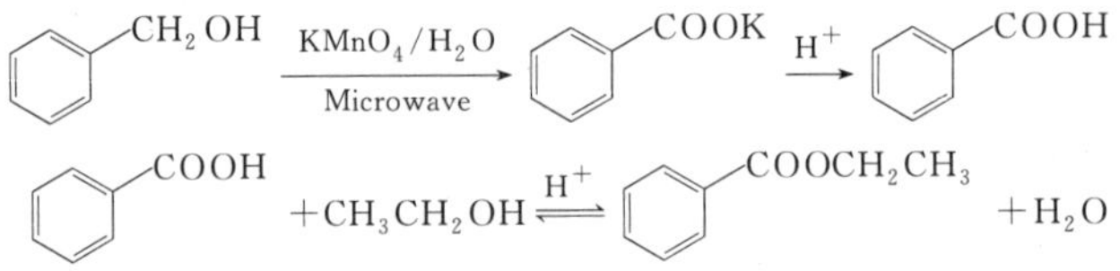

$$\text{PhCH}_2\text{OH} \xrightarrow[\text{Microwave}]{\text{KMnO}_4/\text{H}_2\text{O}} \text{PhCOOK} \xrightarrow{\text{H}^+} \text{PhCOOH}$$

$$\text{PhCOOH} + \text{CH}_3\text{CH}_2\text{OH} \xrightleftharpoons{\text{H}^+} \text{PhCOOCH}_2\text{CH}_3 + \text{H}_2\text{O}$$

Fischer esterification of a carboxylic acid and alcohol is the most common synthetic method for the preparation of ester. Carboxylic acids are not reactive enough to undergo nucleophilic addition directly with weakly nucleophilic alcohos, therefore, the mineral acid is necessary to catalyze the esterification reaction by protonating the oxygen atom of the carbonyl group, making it more electrophilic and thus more conducive to the nucleophilic attack by the alcohol.

Fischer esterification is an equilibrium reaction, and the acid-catalyzed hydrolysis of an ester to the corresponding acid is simply the reverse of the esterification. The equilibrium between the starting materials and products is readily established by heating the reaction mixture under reflux. In order to drive the equilibrium toward completion and increase the yield of the ethyl benzoate, inexpensive and volatile reactant ethanol is used in excess. Meanwhile, the product water is removed from the reaction mixture by azeotropic distillation as it is formed during the esterification.

In this experiment, the benzoic acid prepared by the oxidation of benzyl alcohol is then used directly as the reactant for the synthesis of ethyl benzoate, thus a series of comprehensive experiments are designed together by using the product from the previous step to carry out the following reaction. Such a design meets the requirements of green chemistry, thereby realizing the recycle of experiment products and greatly reducing the generation of laboratory waste.

- **Techniques Involved**

1. Microwave-assisted reflux.
2. Vacuum filtration and washing of solid compounds.
3. Reflux and azeotropic distillation.
4. Extracting, washing and drying of liquid compounds.
5. Vacuum distillation.

Chapter 6 Study Guide of Some Typical Experiments

- **Safety Alerts**

1. Use the microwave oven properly and carefully following the teacher's instructions to avoid the explosions, and be careful while removing vials from the oven as they may be extremely hot.

2. Make sure that the glassware used for microwave-assisted reaction does not have any cracks or chips, otherwise it may shatter during the reaction.

3. Potassium permanganate is a strong oxidant. Avoid contact with it as it will stain skin and clothing. Wash the affected area thoroughly with soap and water.

4. Concentrated hydrochloric acid fume is suffocate and corrosive, so handle the HCl in the hood. If any acid is spilled on your skin, wash it off with large amount of water and then with dilute sodium bicarbonate solution.

5. Concentrated sulfuric acid is corrosive and can cause severe burns. Be very careful while handling it. If any concentrated sulfuric acid comes in contact with your skin, immediately wash it off with copious amounts of cold water and then with dilute sodium bicarbonate solution.

- **Pre-lab Assignment**

1. Try to summarize the advantages and disadvantages of microwave-assisted reactions.

2. In the esterification of benzoic acid, what measures should be taken to improve the yield of ethyl benzoate?

3. For Fischer esterification, it is crucial to remove the newly formed water to drive the equilibrium to the right. What method is hereby employed to remove water as it is formed in this experiment? Why is cyclohexane added to the reaction mixture?

4. In the esterification of benzoic acid with ethanol, the reaction mixture should be heated gently. What is the unexpected effect of flash heating?

5. In the esterification of benzoic acid to ethyl benzoate, the liquid level in water separator should always be controlled, so that the middle layer will not flow back to the flask, and the top layer of liquid is always a thin layer. Explain the reason.

6. What is the purpose of following each operation in the work-up procedures of the esterification of benzoic acid with ethanol?

(1) Extract the aqueous layer with ether(10 mL×2).

(2) Wash the organic layer with 20 mL 10% sodium carbonate.

(3) Wash the organic layer with 5 mL water.

7. What is the principle of vacuum distillation? What does a vacuum distillation apparatus consist of?

8. Try to summarize the general procedure for vacuum distillation.

9. Draft the flow schemes of the oxidation of benzyl alcohol and the esterification of benzoic acid and corresponding work-up procedure respectively for this experiment.

- **Thinking Questions**

1. What is the difference between a vacuum distillation apparatus and a simple distil-

lation unit?

2. Explain the reason why the low boiling point solvents must be distilled out prior to the final vacuum distillation?

3. Write down the stepwise mechanism for the acid-catalyzed esterification of benzoic acid to give ethyl benzoate, and use curved arrows to symbolize the flow of electrons.

4. Concentrated sulfuric acid is used to catalyze the esterification of benzoic acid. Which two steps in the reaction are accelerated by the presence of sulfuric acid according to the reaction mechanism, and what role does sulfuric acid play in each of these steps?

5. Consider the equilibrium for the esterification of benzoic acid to give ethyl benzoate. Provide two specific ways that may be applicable to this equilibrium to drive the reaction completely to the ester.

6. In the esterification of benzoic acid, can the water be distilled out directly after it is formed during the initial step of the esterification, just like the method adopted in the preparation of cyclohexene?

7. The mixture solution of petroleum ether and ethyl acetate with the ratio 4 : 1 can be used as the eluting solvent to monitor the esterification of benzoic acid by thin-layer chromatography. After the reaction mixture is heated for 45 min, apply the reaction mixture on the plate and develop the TLC plate in the chamber. Two spots can be observed on the TLC plate (as the following). What components do the spots A and B represent respectively? If the eluting solvent is changed into a mixture solution of petroleum ether and ethyl acetate with the ratio of 2 : 1, how will the R_f value of spot A change? Explain the reason.

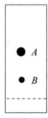

8. The following is the ^{1}H NMR spectroscopy of ethyl benzoate. Try to interpret the spectroscopy and indicate the type of hydrogen that the following absorption peaks correspond to.

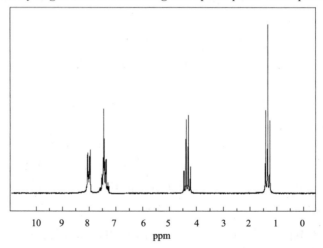

9. Discuss the differences observed in the IR spectra of benzoic acid and ethyl benzoate that are consistent with the formation of the latter in the esterification of benzoic acid with ethanol.

Exp.9 Preparation of Benzoic Acid and Benzyl Alcohol

• **Introduction**

Aldehydes with no α-H, such as aromatic aldehyde, formaldehyde, and trisubstituted acetaldehyde, can undergo mutual oxidation and reduction in the presence of strong alkali. The based-catalyzed mutual oxidation-reduction reaction is also known as Cannizzaro reaction, which results in the formation of carboxylic acid plus alcohols as the oxidation product and reduction product respectively at the same time. Concentrated aqueous solutions of sodium hydroxide or potassium hydroxide are frequently used to provide a strong basic reaction medium, and the reaction scheme studied in this experiment is demonstrated as below:

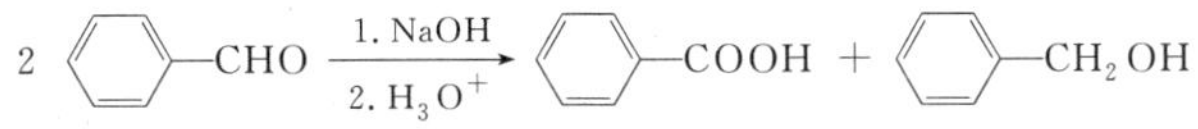

$$2 \text{ Ph-CHO} \xrightarrow[\text{2. H}_3\text{O}^+]{\text{1. NaOH}} \text{Ph-COOH} + \text{Ph-CH}_2\text{OH}$$

In this reaction, benzaldehyde, an aromatic aldehyde, yields benzyl alcohol plus sodium benzoate when heated with aqueous NaOH, and benzoic acid is then obtained by acidification of the aqueous solution of sodium benzoate. This disproportionation reaction takes place by nucleophilic addition of OH^- to benzaldehyde to give a tetrahedral intermediate, which expels hydride ion as a leaving group and is thereby oxidized. A second benzaldehyde molecule accepts the hydride ion in the second nucleophilic addition step and is thereby reduced. The mechanism of the disproportionation of benzaldehyde is shown as follows:

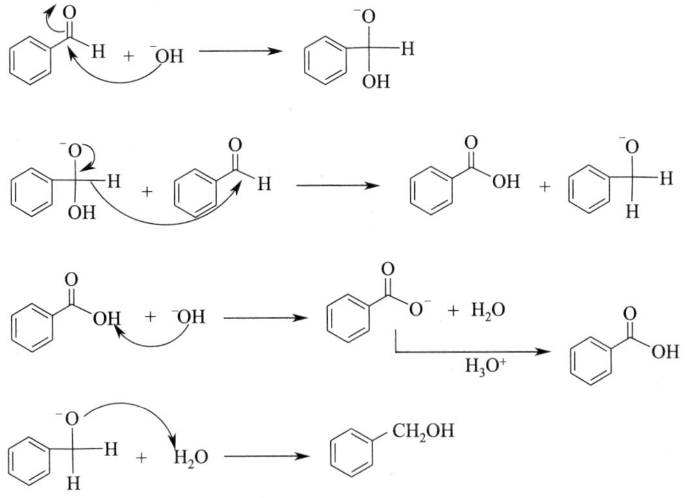

The poor solubility of aldehydes in the concentrated aqueous solutions of alkalies makes the Cannizzaro reaction of benzaldehyde a heterogeneous reaction. In order to facilitate this reaction, vigorous stirring is necessarily important to mix the reactants thoroughly and accelerate the reaction. Meanwhile, high temperature is also involved in the completion of the reaction.

● **Techniques Involved**

1. Reflux and mechanical stirring.

2. Extracting, washing and drying of liquid compounds.

3. Simple distillation.

4. Acidification and vacuum filtration.

5. Recrystallization.

● **Safety Alerts**

1. The concentrated solution of sodium hydroxide is highly corrosive and caustic. Wear latex gloves while preparing and transferring solutions of sodium hydroxide. Do not allow it to contact with your skin. If it does, flood the affected area immediately with water and then thoroughly rinse it with 1% acetic acid.

2. Benzadehyde is toxic and irritating, so wear gloves and avoid skin contact while handling it. Besides, benzadehyde will be easily oxidized to benzoic acid and should be redistilled before use.

● **Pre-lab Assignment**

1. Cannizzaro reaction occurs much more slowly in dilute than in concentrated sodium hydroxide solution. Explain the reason.

2. In addition to monitoring the reaction by TLC, what other methods are available to determine the end point of the disproportionation of benzaldehyde?

3. There are two products of the disproportionation of benzaldehyde, i. e., benzoic acid and benzyl alcohol. What is the principle of the separation and purification of these two products in the experiment?

4. The concentrated hydrochloric acid is added to acidify the aqueous layer from the extraction, converting the sodium benzoate to benzoic acid. Whether is it appropriate to acidify the aqueous solution to neutral by concentrated hydrochloric acid? Explain the reason.

5. Upon the completion of the Cannizzaro reaction of benzaldehyde, the mixture is transferred into a separatory funnel and extracted with ether. Write down the structures of the products dissolved in the organic and aqueous layers, respectively.

6. Saturated sodium bisulfite solution is used to remove the unreacted benzaldehyde from the reaction mixture. Write down the corresponding equation for this chemical change.

7. What is the purpose of each extracting or washing operation in the work-up procedure of the preparation of benzoic acid?

(1) Extract the aqueous layer with ether(10 mL×3).

(2) Wash the ether layer with saturated sodium bisulfite solution.

(3) Wash the ether layer with 10% Na_2CO_3.

(4) Wash the ether layer with H_2O.

8. The crude benzyl alcohol is dried with anhydrous sodium sulfate prior to the final distillation. Can anhydrous calcium chloride be used for this purpose? Explain the reason.

9. Draft flow schemes of the Cannizzaro reaction of benzaldehyde and the corresponding

work-up procedure for this experiment.

• **Thinking Questions**

1. The reaction mixture is always clear after the completion of the initial step of Cannizzaro reaction of benzaldehyde, but may sometimes turn into turbid again when some cold water is added to the mixture. Explain the reason.

2. In the work-up procedures of this experiment, when the ether layer is washed with saturated sodium bisulfite solution, some solids may appear in the separatory funnel. Do these solids need to be filtered out before the next step of washing with 10% sodium carbonate solution? Explain the reason and write down the corresponding equations.

3. In the recrystallization of crude benzoic acid using water, a strong irritant gas is produced when the aqueous solution is heated to the boiling state. Explain the reason.

4. After the benzyl alcohol is collected at its boiling point in the final distillation, some yellowish oil remains in the distillation flask and is rather difficult to distill out. What may this yellowish oily substance be? Write down the corresponding equation for the formation of the oily substance.

5. When the Cannizzaro reaction is performed on benzaldehyde in D_2O solution, no deuterium is found on the benzylic carbon atom in the formed benzyl alcohol. Explain the reason according the mechanism of Cannizzaro reaction.

6. A based-catalyzed crossed Cannizzaro reaction can be performed with two different aldehydes if neither of the aldehydes contains α-H. Write down the stepwise mechanism for the following reaction, and use curved arrows to symbolize the flow of electrons.

$$\text{Ph-CHO} + \text{HCHO} \xrightarrow[\text{2. } H_3O^+]{\text{1. NaOH}} \text{Ph-CH}_2\text{OH} + \text{HCOOH}$$

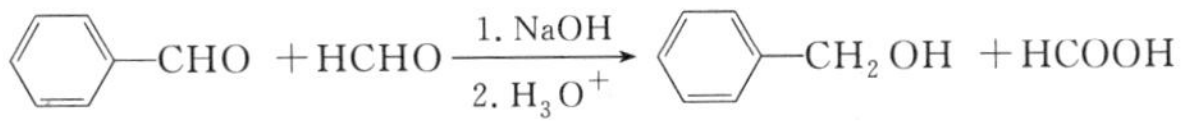

7. Discuss the differences observed in the IR spectra of benzaldehyde, benzyl alcohol, and benzoic acid that are consistent with the formation of the two products from benzaldehyde by the Cannizzaro reaction.

8. Nicotinamide adenine dinucleotide, i. e. , NADH, in biological systems, can reduce carbonyl compounds by a mechanism related to the Cannizzaro reaction. Suppose the mechanism of the following reaction, and use curved arrows to symbolize the flow of electrons.

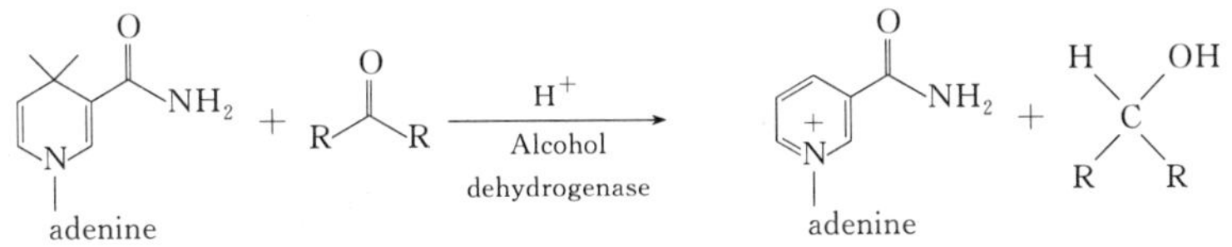

Appendix

Table 1 List of physical properties of some organic compounds

Compounds	M_r	Melting Point /°C	Boiling Point /°C	Density /(g/mL)	Solubility water	Solubility ethanol	Solubility ethyl ether
hexane	86.18	−95	68.95	0.6603	insoluble	50^{33}	soluble
1-butene	56.12	−185.4	−6.3	0.5951	insoluble	soluble	soluble
cyclohexene	82.15	−103.50	82.98	0.8098	insoluble	∞	∞
toluene	92	−92	122.4	0.867	0.047	insoluble	∞
benzene	78.11	5.5	80.1	0.8787^{15}	slightly soluble	∞	∞
p-nitrotoluene	137.14	54.5	238.3	1.1038^{75}	0.004^{15}	—	80.8^{15}
m-nitrotoluene	168.11	90.02	167^{14}	1.574^{18}	0.3^{99}	3.3	soluble
xylene	106.17	—	137-140	0.865	slightly soluble	soluble	soluble
naphthalene	128.19	80.55	218.0	1.0253	0.008^{25}	$9.5^{19.5}$	soluble
1-bromobutane	137.03	−112.4	101.6	1.2758	0.06^{16}	∞	∞
1-chlorobutane	92.57	−122.8	78.6	0.88648^{20}	slightly soluble	∞	∞
bromoethane	108.97	−118.6	38.4	1.4604	1.06^{0} 0.9^{30}	∞	∞
dichloromethane	84.93	−95	40.1	1.3266	slightly soluble	∞	∞
chloroform	119.38	−63.5	61.2	1.485	slightly soluble	∞	∞
bromobenzene	157.01	−30.6	156.2	1.4950	insoluble	∞	∞
chlorobenzene	112.56	−45	131-132	1.107	insoluble	soluble	soluble
benzyl chloride	126.58	−43-−48	179	1.100^{20}	insoluble	∞	∞
benzyl bromide	171.04	−3.9	201	1.4380	insoluble	soluble	soluble
methanol	32.04	−97.8	64.65	0.7915	∞	∞	∞
ethanol	46.07	−117.3	78.5	0.7893	∞	∞	∞
butanol	74.12	−89.53	117.25	0.8098	9^{15}	∞	∞
cyclohexanol	100.16	25.15	161.5	0.962	3.6	soluble	soluble
2-methyl-2-butanol	88.15	−8.4	102	0.8059	∞	∞	∞
furfuryl alcohol	98.10	—	170-171	1.1282	∞	soluble	soluble
tert-butanol	74.12	25.5	82.5	0.7858	soluble	∞	∞
benzyl alcohol	108.15	−15.3	205.35	1.0419	4^{17}	soluble	soluble
triphenylmethanol	260.34	164.2	380	1.199^{0}	insoluble	soluble	soluble
phenol	94.11	40.85	182	1.071^{41}	soluble	soluble	soluble
hydroquinone	110.11	172	285/730mmHg	1.332	soluble	soluble	soluble
o-tert-butylhydroquinone	166.22	127-129	—	—	soluble(hot)	soluble	soluble
diethyl ether	74.12	−116.2	34.51	0.7138	7.5	∞	∞
di-n-butyl ether	130.23	−95.3	142	0.7689	slightly soluble	∞	∞
benzaldehyde	106.13	−26	178	1.0415	0.3	∞	∞
acetone	58.03	−95.35	56.2	0.7899	∞	∞	∞

Continued

Compounds	M_r	Melting Point /℃	Boiling Point /℃	Density /(g/mL)	Solubility water	Solubility ethanol	Solubility ethyl ether
benzophenone	182.22	48.5	305.4	1.0869^{50}	insoluble	soluble	soluble
acetophenone	120.15	20.5	202.6	1.0281	slightly soluble	soluble	soluble
benzoin	212.22	137	344	1.31	soluble(hot)	soluble(hot)	slightly soluble
benzil	210.22	95	346-348	1.23	insoluble	soluble	soluble
acetic acid	60.05	16.6	117.9	1.049	∞	∞	∞
cinnamic acid	148.17	133	300	1.2475	0.04^{18}	24	soluble
benzoic acid	122.12	122.4	249.2	1.266	0.2^{17}	46.6^{15}	66^{15}
phenylacetic acid	136.16	77	265.5	1.0914^{77}	slightly soluble	soluble	soluble
furoic acid	112.09	133-134	230-232		soluble	soluble	soluble
chlorosulfonic acid	116.53	−80	151-152 (decompose)	1.753	decompose	decompose	—
p-toluenesulfonic acid	172.20	106-107	140/20 mmHg	—	soluble	soluble	soluble
o-aminobenzoic acid	137.14	144-146	—	1.412	soluble(hot)	soluble	soluble
acetylsalicylic acid	180.15	135	—	—	0.333	20	6.7~10
salicylic acid	138.12	159	211/20 mmHg	—	slightly soluble	37	33.3
glycollic acid	76.05	80	—	—	soluble	soluble	soluble
acetic anhydride	102.09	−73	140	1.082	12	soluble	insoluble
phthalic anhydride	148.12	130.8	295	1.53	soluble	soluble	—
ethyl acetate	88.12	−83.57	77.06	0.9003	8.5^{15}	∞	∞
n-butyl acetate	116.16	−77.9	126.5	0.8825	0.7	∞	∞
ethyl benzoate	150.18	−34.6	213	1.0468	insoluble	soluble	∞
ethyl acetoacetate	130.14	−80 (enol form) −39 (ketone form)	180.8	1.0250	slightly soluble	soluble	soluble
benzyl acetate	150.18	−51	213	1.057	insoluble	∞	∞
benzoyl chloride	140.57	−0.5	197.2	1.2188	—	—	soluble
p-acetamidobenzene sulfonyl chloride	233.68	149	—	—	decompose	soluble	soluble
dimethylformamide	73.10	−61	152.8	0.9487	∞	∞	∞
phthalimide	147.13	238	—	—	slightly soluble	—	—
sulfanilamide	172.21	164.5-166.5	—	—	slightly soluble	slightly soluble	insoluble
N,N'-dimethylaniline	121.18	2.5	192.5-193.5	0.956	slightly soluble	soluble	soluble
aniline	93.13	−6.3	184.13	1.0213	3.6^{18}	∞	∞
acetanilide	135.17	114.3	304	1.219^{15}	0.53^6 5.2^{100}	21	7^{25}
m-nitroaniline	138.13	114	305-307 (decompose)	1.43	slightly soluble	soluble	soluble
hydrazine hydrate	50.06	−51.7 (decompose)	119.4	1.032	∞	∞	insoluble
methyl orange	269.30	181-182	—	—	insoluble	soluble	—
benzpyrole	117.15	52.5	254	1.22	soluble	soluble	soluble
furfural	96.09	−38.7	161.7	1.1594	soluble	soluble	soluble
tetrahydrofuran	72.11	−108.5	67	0.8892	∞	∞	∞
pyridine	79.10	−42	115-116	0.9780	∞	∞	∞
α-aminopyridine	94.12	58.1	204 (sublimation)	—	soluble	soluble	soluble

Note: Solubility repers to the number of grams that a substance dissolves when reaches saturation in 100 mL solvent at a certain temperature. The superscript numbers in this table refer to the temperature(℃) at the time of measurement.

Table 2 List of physical properties for some commonly used solvents

Solvent	Boiling point /℃	Density /(g/mL)	Miscible with water	Solvent	Boiling point /℃	Density /(g/mL)	Miscible with water
methanol	64.96	0.791	√	benzene	80.1	0.879	×
ethanol	78.5	0.789	√	toluene	110.6	0.867	×
butanol	117.25	0.810	slightly	o-xylene	144	0.897	×
ethyl ether	34.51	0.714	×	m-xylene	139.1	0.868	×
acetone	56.2	0.790	√	p-xylene	138.4	0.861	×
acetic acid	117.9	1.050	√	chloroform	61.7	1.483	×
ethyl acetate	77.1	0.900	slightly	dichloromethane	40	1.326	×
tetrahydrofuran	66	0.890	√	CCl_4	76.54	1.594	×
acetonitrile	81.65	0.790	√	DMF	153.0	0.940	√
cyclohexane	80.7	0.780	×	DMSO	189.0	1.10	√
hexane	69.0	0.660	×	NMP	204.0	1.03	√

Table 3 List of some elements with their symbols and atomic masses

Element	Symbol	Atomic number	Atomic mass	Element	Symbol	Atomic number	Atomic mass
silver	Ag	47	107.868	iodine	I	53	126.905
aluminum	Al	13	26.9815	potassium	K	19	39.089
bromine	Br	35	79.904	magnesium	Mg	12	24.305
carbon	C	6	12.011	manganese	Mn	25	54.9380
calcium	Ca	20	40.078	nitrogen	N	7	14.0067
chlorine	Cl	17	35.453	sodium	Na	11	22.9898
chromium	Cr	24	51.996	oxygen	O	8	15.9994
copper	Cu	29	63.546	phosphorous	P	15	30.9737
fluorine	F	9	18.9984	lead	Pb	82	207.2
iron	Fe	26	55.847	sulfur	S	16	32.06
hydrogen	H	1	1.0079	tin	Sn	50	118.69
mercury	Hg	80	200.59	zinc	Zn	30	65.38

Table 4 List of the azeotropic boiling points of some azeotropes

Azeotrope components	Percentage by weight/%	Azeotropic boiling points/℃	Azeotrope components	Percentage by weight/%	Azeotropic boiling points/℃
PhMe-H_2O	79.8 : 20.2	79.8	EtOAc-H_2O	91.8 : 8.2	70.4
PhH-H_2O	91.1 : 8.9	69.3	PyH-H_2O	57.0 : 43.0	92.6
BuOH-H_2O	62.0 : 38.0	92.4	MeCN-H_2O	83.7 : 16.3	76.5
iso-BuOH-H_2O	66.8 : 33.2	90	cyclohexane-H_2O	58.4 : 41.6	70.8
sec-BuOH-H_2O	67.9 : 32.1	88.5	$CHCl_3$-H_2O	97.0 : 3.0	53.3
tert-BuOH-H_2O	88.3 : 11.7	79.9	EtOH-cyclohexane-H_2O	9.0 : 7.8 : 76	62.1
cyclohexanol-H_2O	58.3 : 41.7	97.8	EtOH-PhH-H_2O	18.5 : 7.4 : 74.1	64.9
cyclohexanol-cyclohexene	69.5 : 30.5	64.9	EtOH-PhMe-H_2O	37.0 : 51.6 : 11.4	74.4
2-methyl-2-butanol	72.5 : 27.5	87.4	EtOH-EtOAc-H_2O	9.0 : 83.2 : 7.8	70.0
THF-H_2O	95.0 : 5.0	65	EtOH-$CHCl_3$-H_2O	4.0 : 92.5 : 3.5	55.5
PhOMe-H_2O	59.5 : 40.5	59.5	BuOH-Bu_2O-H_2O	34.6 : 34.5 : 29.9	90.6
Bu_2O-H_2O	33.4 : 66.6	94.1			

References

1. Lan Zhou University, Fu Dan University. *Experimental Organic Chemistry*. 2nd ed. Beijing: Higher Education Press, 1994.
2. Zeng Z Q, Zeng H P. *Experimental Organic Chemistry*. 3rd ed. Beijing: Higher Education Press, 2001.
3. Allen M S, Barbara A G, Melvin L D. *Microscale and Miniscale Organic Chemistry Laboratory Experiments*. 2nd ed. New York: McGraw-Hill Companies Press, 2004.
4. Xue S J, Ji P. *Experimental Organic Chemistry*. 2nd ed. Beijing: Science Press, 2007.
5. Jerry R M, Christina N H, Paul F S. *Techniques in Organic Chemistry*. 3rd ed. New York: W. H. Freeman and Company Press, 2010.
6. John C G, Stephen F M. *Experimental Organic Chemistry: A Miniscale and Microscale Approach*. 5th ed. Boston: Cengage Learning Press, 2011.
7. Andrew P D. *Green Organic Chemistry in Lecture and Laboratory*. Boca Raton: CPC Press, 2012.
8. Wu X Y, Wang Z, et al. *Experiments in Organic Chemistry*. Beijing: Tsinghua University Press, 2012.
9. Zhang P F, Qiang G R, Zhao H R. *Organic Chemistry Experiments*. Hangzhou: Zhejiang University Press, 2013.
10. Qiang G R, Jin H W, Sheng W J. *Organic Chemistry Experiments*. 3rd ed. Beijing: Chemical Industry Press, 2020.
11. Li Y, Shao Y. *Laboratory Experiments for Organic Chemistry*. Nanjing: Nanjing University Press, 2021.